Seven-Tenths

Seven-Tenths

Love, Piracy, and Science at Sea

David Fisichella

A LeapSci Book
Leapfrog Science and History
Leapfrog Press
Teaticket, Massachusetts

A LeapSci Book
Leapfrog Science and History

Published in 2010 in the United States by
Leapfrog Press LLC
PO Box 2110
Teaticket, MA 02536
www.leapfrogpress.com

Distributed in the United States by
Consortium Book Sales and Distribution
St. Paul, Minnesota 55114
www.cbsd.com

First Edition

Library of Congress Cataloging-in-Publication Data

Fisichella, David.
Seven-tenths : love, piracy, and science at sea /
David Fisichella.
p. cm.
"A LeapSci Book Leapfrog Science and History."
ISBN 978-1-935248-10-1 (alk. paper)
1. Fisichella, David. 2. Blindness. 3. Oceanography. 4. Oceanographers.
5. Marine sciences–Research. 6. Voyages and travels. 7. Seafaring life.
8. Bower, Amy S. I. Title. II.
Title: Love, piracy, and science at sea.
GC30.F57A3 2010
551.46092'2–dc22 2009052905

Printed in the United States of America

To Amy, for providing me with the opportunity to write this book and for being my personal tutor, both in oceanography and in life.

Acknowledgments

I wish to thank my parents for instilling in me a love of reading, from which all writing comes.

A sincere thank-you to Judy, Nancy, Tara, Ethel, Migdalia, my editor Kimberly Davis and my publisher Lisa Graziano, who were kind enough to share with me their knowledge of writing. This is as much their book as it is mine.

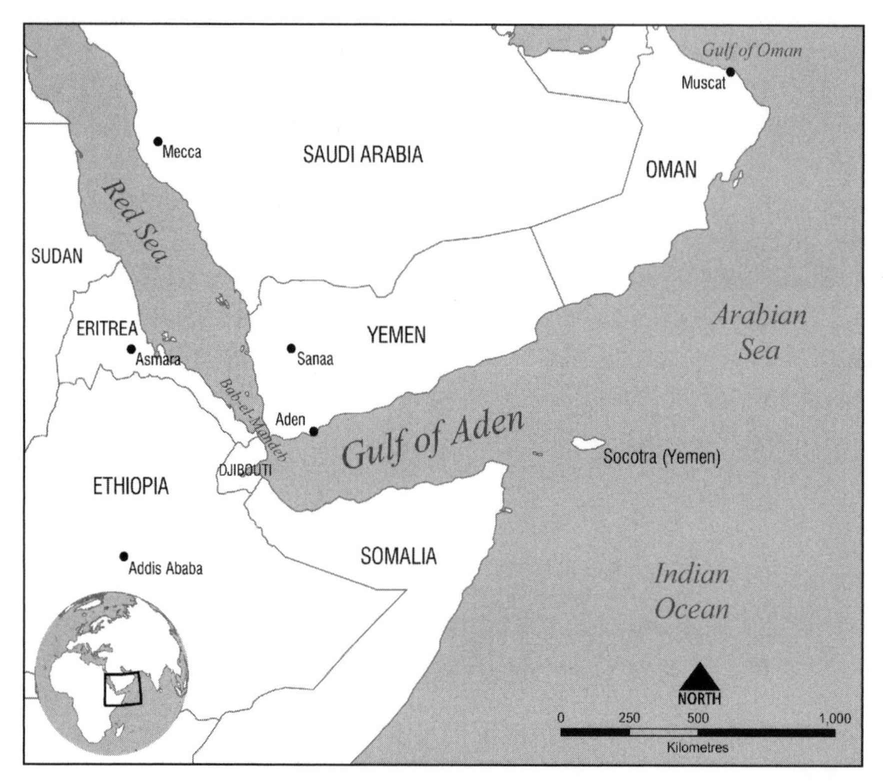

Location of the Gulf of Aden, which joins the Red Sea to the Arabian Sea and the Indian Ocean. The equator is just off the bottom of the figure, crossing through Kenya (south of Ethiopia) and the southern tip of Somalia.

PROLOGUE

My world at the moment was a white porcelain cave with smooth walls and a lake at the bottom. The toilet bowl felt cool under my chin. I had never been so violently ill. But even though my head was in the toilet, my life had been out of one for a couple of years now.

Amy heard the bowl flushing and walked into the bathroom with a glass and a bottle of bubbly water.

"How are you feeling?" she asked, knowing the response would be obvious, but exhibiting a sense of empathy that we both agreed I would never seem to possess.

"I'll make it to the ship as long as we don't get stuck in traffic," I responded, still on my knees. I took the glass from her hand. "How much time before we have to leave?"

"According to the agent the van will be here in two hours." She sat down on the side of the tub.

There was never any real concern over traffic. The only congestion we would likely encounter on the way to the port was a slow-moving herd of goats, typical of Djibouti, an African country pretty low on the list of car ownership per capita.

Amy sat there quietly. This was something else she claimed to do better than I—sit and listen. My need to be proactive in fixing emotional and social problems instead of just listening was a deficiency of male behavior that she would point out frequently. This time, however, I think she was not so much waiting to sympathize with my febrile condition as contemplating the impact my illness would have on the start of her research cruise, only hours away.

She was wearing a light cotton tank top and shorts, her hair held back from her face by a clip. This was less about fashion and more an attempt to mitigate the oppressive August heat that permeated the stained cinderblock walls of the hotel. The air conditioning had lost its battle with the North African summer long ago. I was reluctant to get off the tile floor, which was acting as a heat sink for my already elevated body temperature.

"Would you feel better if you had something to eat?" Amy asked.

I stood up slowly. "Eating is probably what got me into this mess," I said.

She handed me two packages of granola snack mix. "Can you tell me which one has raisins? I know how much you hate them, so I'll eat that bag," she said. Amy had a vision problem that prevented her from seeing things that small, so I had to read the labels. It was in fact the only reason I was there—to assist her with her research by being a good set of eyes. At the moment, however, she was providing most of the assistance.

The room seemed to be spinning in the opposite direction of my stomach, but as bad as things were, I found myself smiling. Two years ago I would never have imagined myself throwing up in a moldy bathroom in an obscure African country while preparing to go on a scientific expedition. I stood in front of the cracked mirror and looked at my now-gaunt face, proof that the most interesting adventures in life cannot be planned.

So how did an out of work engineer with no deep-sea experience, whose only exposure to oceanography was watching the Discovery Channel, find himself heading for sea as a member of a scientific party from a prestigious oceanographic institution? It all started with a blind man and a sailboat.

1

IT WAS SOMETIME PRE-1992. I can't remember the exact date, but it was spring and everything outdoors was covered in a fine coating of yellow pollen. The setting sun projected through the hotel room window, casting orange shadows from the furniture onto a painting of flying ducks on the wall. The glowing hues made the birds look as if they were fleeing an inferno of marsh grass. It had not been a good day. More accurately, it had been a typical day, which for the last few years meant that it was not so good. I cannot say I had gotten used to this, which was promising. If I had become complacent with life under a constant fog of ambivalence, I probably would not have done anything to change the situation. Complacency, apathy, routine—only words, but killer-of-men words. I wasn't ready to die, but I also wasn't doing a very good job of living either. I was existing, quite well by most standards; I had my own home overlooking Boston Harbor, a career for which I was well-compensated, and a boat. Hey, things can't be that bad if you have a boat, I often reminded myself.

The boat was a refuge. She was the center of my social life. The marina was the equivalent of the neighborhood pub, a place for gathering, drinking, and talking about trivial things that all seemed very important at the time. The boat also had the qualities of a good spouse: you took care of it and it took care of you. Certainly, there were times when I had to take head plumbing apart and stare into pipes coated with years of excrement, but the rewards of chasing a rising moon on a balmy summer night made up for those times deep in shit.

Lying on the bed and staring at the burning ducks made me realize that lately I was spending more time in shit than chasing the moon. The boat was 3000 miles away and I had just wasted four days of my life trying to convince a small business owner in California to spend more money than he could afford to remake a batch of tiny, inconsequential widgets that were a few thousandths of an inch out of specification, even though we both knew the tolerance was overly tight. It's too bad he couldn't have negotiated a cost-plus contract with us, as my company had done with the Defense Department. That way he would have been paid to make mistakes, just like we were. It felt dirty to be a part of it all, but complacency, apathy and routine, words never far away, were there to hide the grime. I had spent the last two years on trips like this—hotel rooms with duck paintings, companies housed in buildings of beige concrete situated in industrial complexes ringed with TGIFridays, Denny's, and Olive Garden restaurants. Without the automobile license plates it would have been impossible to tell what state I was in.

The company I worked for developed and manufactured weapon systems for the military. We fired missiles, blew up tanks, and manufactured RADAR to land jets on the decks of pitching aircraft carriers. It sounded exotic, but after a dozen years, the business had become distilled to the mundane production of many hundreds of small parts. The projects I worked on were in development for many years, if not decades, and were often ended abruptly when defense budgets were reallocated. A look back on my tenure there, and seeing thousands of hours of my labor made meaningless when a project was terminated based on the shifting political wind, only fueled the disillusionment.

I looked at the clock. I should have been writing my report of what I had found, but I couldn't bring myself to put the words on the paper. I didn't know what to write. As an engineer who looked at manufacturing processes, I could summarize most of what can go wrong in two dozen standard phrases: "The procedure was correct, but wasn't followed"; "raw material not within limits"; "operator error"; the list went on. In the beginning, it was fun, like a puzzle. Figure out what went wrong and then fix it. Somewhere along the line the job lost its challenge, then its meaning, and as I lay there in that hotel room I came to the realization that I wasn't very good at my work anymore.

The thought of picking up a pen and reaffirming this revelation wasn't appealing. At this moment I needed to do something that didn't remind me of failure, either mine or someone else's. Since I didn't have my boat to retreat to, I settled for a sailing magazine.

After an hour of reading about the adventures of others sailing in and out of exotic ports I came across a sidebar article. Only a few paragraphs long, it read more like an advertisement. The Carroll Center for the Blind in Newton, Massachusetts, was looking for volunteers to sail as sighted guides for the blind. I read the section quickly and moved back to the main article about the vagaries of clearing customs on some remote pacific atoll, but my mind kept wandering back to blind sailing. How would a bunch of blind people navigate around the harbor without running into something, or make sense of all the lines that typically cluttered a boat's cockpit?

I kept turning back to those few vague paragraphs in search of clues, but there was nothing to be found except a phone number. I looked at my watch. It was 7:50 p.m., not quite 5:00 back in Newton. I picked up the phone and caught Arthur O'Neil as he was walking out of his office. Arthur, it turned out, was the director for the Carroll Center's Outdoor Enrichment Program and organized the blind sailing program. "So how does this work?" I asked.

"First, I'll determine your sailing abilities. Then pair you up with a team. The guys on your team will teach you the proper guiding techniques," he said, sounding a bit rushed to get home for the evening.

"I've done plenty of sailing, but never with a blind person," I said.

"Don't worry, there'll be plenty of time for practice before the first regatta next month."

"What do mean 'regatta'? Blind people will be racing these boats?" I asked incredulously.

"Yeah, it's a lot of fun, really. Come down and give it a try. I think you'll like it. I gotta run. We start next Saturday at nine at the Courageous Sailing Center. I hope to see you there."

I managed to get some cryptic directions from him before he hung up, and with more skepticism than enthusiasm I told him I would be home by then and would check it out.

My flight returned to Logan Airport early the next day. It was a red-eye from LAX, and as I stood looking into the mirror in the men's room at the terminal, I knew why. The puffy, bloodshot orbs that looked back at me confirmed what I already knew. It was time for a change, but to what?

The cab ride to the house took all of 10 minutes. I opened the front door and dropped my small suitcase on the floor. A calico cat lay in a sunbeam, stretched out sideways as if in mid-stride. She was fat and didn't stride much these days, and in keeping with her sedentary lifestyle, didn't even lift her head when I walked by. The other cat, the small black one, was nowhere to be seen. She was probably perched in her favorite window, stalking birds she could only dream of catching.

My wife, Lisa, was also not around. She was working, or more probably shopping, since she only worked sporadically. The not-working part didn't bother me, but the spending part did. It was becoming a serious issue in our relationship, but not the issue that was sending us over the precipice. What was dissolving our 10-year marriage was suicide: Lisa's. Lisa was diagnosed as an insulin dependent diabetic a few months before our wedding and in the 3650 days since then she had never admitted to herself the seriousness of her condition. Lisa saw insulin as a hormone that made her gain weight, not one that kept her alive. The trajectory of all our lives moves in the same direction—toward death. Lisa's problem was that she was determined to get there more quickly than I was. This attitude could lead to only one outcome. I was watching her kill herself ever so slowly by not taking her doses and not monitoring her blood sugar. I had spent a decade cajoling, bribing, yelling, whatever it took to get her to take care of herself, but nothing I said or did made any difference. Any talk of children was out of the question. Lisa was in no condition to bear children, and even if she was, I was not going to bring children into this world only to have them see their mother die at an early age. This, I told myself, was yet another of my failures. I didn't have what it took to help her or find someone who could. Being willing to help and being able to help didn't seem to connect. For whatever reason, I had failed her. I was quitting, and felt guilty for it. I was out of ideas and the only thing left to me was anger.

In my marriage, as in my job, I knew it was time for a change, but didn't know to what. Where I was wasn't good, but at least it was predictable. This was another warning sign: I was in a bad situation and justified it because it was familiar. I was one step closer to the abyss. I looked out the window to the boats resting on their moorings. Tomorrow was Saturday, the day I was going to meet up with Arthur O'Neil. The idea of being out on the water with a boat full of blind sailors was suddenly very appealing.

The next morning I got in the car with some trepidation about what I was about to do, but also with a sense of adventure I had not felt in a long time. My image of the blind was personified by Ray Charles. Blind people either played pianos in smoke-filled nightclubs or ran concession stands in federal buildings. As I drove to the marina, I wondered how they would fit the baby grand in the boat.

At the wharf I found the entrance to the Courageous Sailing Center. Appropriately named, I thought. I didn't know what would take more courage, getting into a boat with a blind guy at the helm or being the blind guy and having to rely on me to tell him where to steer. Arthur must have identified me by the apprehension on my face. He came up and introduced himself. "Welcome to the SailBlind program," he said, his wide smile radiating from under the brim of a faded ball cap bearing the name of a regatta sailed years ago. Arthur showed me around the facility and introduced me to a number of guides and blind sailors. I was so overwhelmed I forgot each one's name before the handshake was over.

We continued our progress down the pier. At the end, leaning against a wooden pile, was a man with a cane. He was in his forties, with a worn pair of Sperry Topsiders on his sockless feet and a pink polo shirt neatly tucked into khaki shorts. His face was turned up to catch the warm rays of the sun, which, even this early in the summer, had already freckled his pale skin. Arthur introduced him as Doug, one of the people I would be sailing with. The three of us chatted for a few moments before Arthur excused himself to attend to something.

"So, where do you live?" Doug asked.

"Winthrop. It's on the other side of the harbor."

"No kidding?" he said. "I moved there a few months ago."

I realized this may have been part of Arthur's plan in pairing us up. It wasn't very easy to get to Courageous from Winthrop without a lot of bus and train connections. He probably figured we would drive together. That was fine with me, but I was hoping he paired up his teams using more important criteria than geography.

"How long have you been sailing here?" I asked.

"A few years. It's been great. We've raced in a bunch of places, even went as far as New Zealand for the World Blind Sailing Championships last year."

Arthur hadn't mentioned anything like that. This was beginning to sound like just the thing I was looking for. "What happened to your previous guide?" I asked.

"People come and go," he said without offering any more details. Before I could press him for more information, there was a tap-tap-tap from behind me and another man with a cane joined us.

"Hey Dougie, what's happen'in," he said in a thick Boston dialect that was somewhat slurred. Most of the sound came out his nose in long, drawn vowels. "So you're Dave. Pleased to meet'ya. I'm Don. Art told me you were here." He slapped me on the back in lieu of a handshake. Don was rail thin, with jaundiced features and dark thinning hair combed straight down on all sides of his head. He was wearing a clunky pair of wraparound sunglasses and a tee shirt with the sleeves cut off. I couldn't tell his age, but my guess was he was younger than his features made him look.

Doug pointed his cane in Don's direction. "Watch out for this guy. He likes to joke around at other people's expense."

"Hey, that's not true," Don said.

"Once he put his glass eye into his beer and told the guy next to him at the bar, 'keep an eye on my drink while I'm gone,'" Doug said to me.

"You come up with a good prank once, and look what happens," laughed Don, seeming pleased with his reputation. This was shaping up to be a fun boat.

The three of us spent the next few hours out on the water. The wind was light and the sailing uneventful. Both Doug and Don were very patient with me. Just the act of rigging the sails had me hunting for ways to put into words what I'd always done without thinking, and the words did not always amount to anything intelligible. Once on the water everything began to fall into place. Don sat forward of me and trimmed the jib. I handled the mainsail and Doug focused on handling the tiller. I began to learn how and to what extent Doug and Don needed to have things described. They also taught me the correct commands for giving direction. Being consistent was essential. When the winds picked up things would get hectic, leaving no room for miscommunication. As we drifted around the moored boats in the harbor, I learned a lot about what brought Don and Doug to this place.

Both suffered from diabetes and were blind from retinopathy, a complication of the disease. They told similar stories; blood vessels in their retinas ruptured, destroying the surrounding cells and slowly taking the view of the world with them. As I listened to Doug and Don, I could not help thinking of Lisa suffering the same fate.

There were other parallels. Like Lisa, Doug lived the early part of his life in denial about his diabetes and soon became one of the "angry blind." Doug would stumble through his day with a red-tipped chip on his shoulder, devastated at the loss of independence yet abusive to those who tried to help him. I was angry too. Angry at where I'd let my life go, and listless about doing anything to change it. Doug found an outlet for his anger in racing sailboats. I could relate to that. Racing a winning boat requires concentration and focus. There's no place in the cockpit for anger. The angry skippers are the ones always looking for crew.

Within a short time I learned to become Doug's eyes on the water. I became good at describing our relationship to other boats, looking out for large waves, and guiding him into the dock at the end of the day. As the helmsman, Doug was always in control of the boat, steering by the pressure on the tiller and the wind against his face. He provided feedback to Don and me on how the boat was handling so that we could trim the sails quickly. Doug's greatest challenge was filtering my occasional right-left dyslexia, a condition that had not been noticeable until I needed to tell a blind man which way to move the helm to avoid hitting a buoy.

Don worked hard at trimming the jib. On blustery days he taxed his thin arms trying to bring the sail in tight, always giving everything he had. He also talked a lot. When the wind was light, I swore all that air coming from his mouth helped move us along.

By the end of a few weeks in the Sail Blind program I had learned a lot about blindness. Most of the people on the team who were considered blind had some useful vision. It was hard to find someone in the group who could not detect the difference between day and night. I also got to practice what I knew about diabetes. My unofficial role as guide to two diabetics was to watch for signs of low blood sugar. It was not unusual for me to look aft and see Doug at the helm on the verge of unconsciousness, his blind eyes staring off into space. At those times I would pry his fingers from the tiller and hope that he was alert enough to drink some juice. In the beginning, this felt like a mountainous responsibility, but over time it became one of the things that brought us together.

Eventually we become a true team. When I said "up a little" or "you're pinching" Doug knew exactly how much to change course. Conversation was minimal in the boat during a race. Like a long-married couple, we anticipated what the other person was going to say before he said it.

Doug, Don, and I continued sailing together for the next few years, traveling to regattas around the country and even as far away as Australia. The three of us became good friends, though Doug and I developed a more complex bond. I think it was due to his intensity. Don was carefree, happy with a disability check and a cold beer. Doug wanted more. I could appreciate that.

Doug died suddenly in 1994 just as things in his life were improving. He passed away in his sleep from an insulin reaction. I never told him how much I appreciated his companionship, but I think he knew. People would say, "Isn't it nice what you do for Doug?" They didn't understand. By that time, I wasn't "helping the poor blind guy"; I was sailing with a friend. He never wanted or needed pity from anyone, and, I learned, neither did I. I wanted things to change, and I was the only person capable of setting that change in motion.

When I first met Doug my life was on the edge. Peering over the precipice into a valley of depression, I saw an image of what awaited me if I didn't do something, anything. The solution was adventure. It didn't have to be climbing Everest or anything so dramatic. My adventures needed to be only small challenges to wake me from a day of sleepwalking from task to task. I found this adventure in meeting new people. I found it sailing with a friend whose blindness did not define him in my eyes any more. And I found it at a party, in a most unexpected way.

2

I STARED AT THE INVITATION ON MY DESK. It was for a reception at the Carroll Center for the Blind to highlight a new Outward Bound program the school was involved with. I had not had much contact with the sailing program since Doug's death, but the remote possibility of running into some of the sailors I had known was appealing. I'd just left a meeting in which I'd lost my temper. My job had not improved in recent months and I was feeling new levels of anger and resentment. "This isn't me," I kept telling myself. Of course, it *was* me, or who I had become, and I did not like the realization. At age 33, I was seeing my life going up and down like a seesaw, bouncing between highs and lows and never finding the balance point of happiness. I looked at the invitation again. The other choices were going home to a wife I was in the process of divorcing or staying late at a job that had long since burned me out. Either was more unbearable than the idea of attending a party of people I didn't know, for a cause I wasn't connected with. I grabbed the car keys.

The darkened room at the Carroll Center was filled with a mix of men and women, blind and sighted, in orderly rows of chairs. I was sitting in the back row next to a projector that mechanically regurgitated another slide in front of the lens. The screen showed a half-dozen people, all wearing helmets and harnesses. A lone climber was suspended from a rope halfway up a sheer cliff. The speaker at the front of the room described in great detail what was happening in the pictures. The presentation was good, but for some reason I was not engaged by it. I found myself daydreaming about running my own yacht charter business in the Florida Keys where the sun was warm, the ocean was a translucent blue, and young women vied for a turn to help me steer the boat.

The lights of the room came up, interrupting my wandering thoughts. The fan of the projector continued to hum, blowing warm air onto my arm. From what I'd seen of the presentation, the program had brought together blind participants from the Carroll Center, and managers from a large Boston accounting firm, at Outward Bound's Hurricane Island in Maine. The course was designed to provide a physical challenge for the blind while offering a lesson in team building for the company's employees. I couldn't picture anyone I worked with completing the ropes course, but I liked the idea of a few of my colleagues dangling from the end of a rope.

I stood up and walked over to the bar, picked up a drink, and moved toward someone in the room I recognized, Ted, a member of the sailing team. Judging by the preceding presentation, he was also a graduate of the Outward Bound School. Ted lost his vision as a result of retinitis pigmentosa, a genetic disease that slowly shrinks the field of view until the small tunnel of light finally turns black. Ted was talking to another Outward Bound climber, an attractive young woman. In the photos she was trussed in an ungainly climbing harness and helmet that obscured her features. Here, dressed in black silk slacks, gold blouse and a finely embroidered vest, she exhibited the same confident stance that made her easily recognizable no matter what she wore. She didn't have a cane, and from the presentation it wasn't clear if she was one of the managers or one of the blind participants. I moved into that awkward zone of space at the fringe of their conversation in the hope of being invited in. Ted was polite, though I sensed that I had interrupted his well-prepared plan to coax this woman back to his apartment. After what seemed like an intentionally long delay, he made a stiff introduction. "David, this is Amy Bower," he said.

Amy turned toward me, seeming pleased to have someone else to talk with "So, how long have you been an accountant?" I asked her.

She looked puzzled. "I'm not an accountant, I'm an oceanographer."

At that point I must have been the one with the puzzled look on my face as I tried to process why an accounting firm would need an oceanographer. Short of counting fish, I could not come up with a reason. Ted grinned, but Amy stepped in to curtail my embarrassment.

Smiling, she said, "That's OK, it happens all the time. I'm legally blind, but I get around pretty well."

I had never met an oceanographer. I immediately wanted to find out more about her work. I suppressed the urge to ask all the standard questions that came to mind: "Do you know Jacques Cousteau?"' "Why do whales beach themselves?" "What's it like to swim with dolphins?" and tried to sound interested but still a little aloof, if for no other reason than to get further with her than Ted had.

Amy was a scientist with the Woods Hole Oceanographic Institution on Cape Cod—a place I had only heard about from PBS television documentaries narrated by men with British accents. For the next hour this woman—who despite her vision problem traveled the world oceans, loved outdoor sports, and understood calculus—kept my attention long after I had intended to leave the party. Amy explained that there was a whole field of oceanography that dealt with the physical characteristics of the earth's oceans. Where water circulated, what its properties were, and how it interacted with the air to affect our climate. These were the things that physical oceanographers studied. Amy, it turned out, was a physicist, and knew nothing about marine mammals or French SCUBA divers.

The room was warm with the heat of overdressed people, and the crowd was starting to thin. During our conversation Amy had formed a subtle wedge between Ted and me. Sensing the intent of the body language, Ted decided to cut his losses and strike out for better fishing in another corner. "Sorry if I interrupted something here earlier," I said as he walked off.

"Don't apologize. If anything I should be thanking you," Amy said, cocking her head in Ted's direction.

"Tell me more about your vision problem," I said, wanting to find out how someone who was blind, even if only by some legal definition, could go to sea and do research. Having sailed with Doug and Don I knew that the limitations we placed on persons with disabilities were mostly artificial, but in Amy's case I was curious about how she adapted to her work.

"My problem is macular degeneration, a loss of central vision. I don't see the detail in anything, can't read text unless it's the size of my fist, and they won't let me drive a car. My peripheral vision is not bad though, and I can walk around a room without bumping into too many things."

"How do you do your job?"

"Adaptive devices. I've got a computer that talks to me and a machine that enlarges printed material on a screen. It's not easy, but I get by." The corners of her mouth turned up in another smile.

Here was a woman who faced a lot more challenges than I did in an average day, and she smiled a heck of a lot more than me. That was the moment I became very interested in her.

Amy asked me about my job. She seemed seriously interested in my answers, not just in polite conversation. It turned out she knew Doug, and had done some sailing with the Sail Blind program before I became involved. We talked about sailing and growing up on the North Shore of Massachusetts, me in Winthrop and Amy in Rockport, which were not that far apart. We both admitted that we couldn't live out of sight of the ocean.

Amy told me more details about her research. She spoke with an enthusiasm for her work that made it sound not only interesting, but exciting. It was a feeling I had not experienced in my own job for a long time. Her enthusiasm was infectious. "I'd love to be able to do something like that," I said.

"Do you get seasick?" she asked.

I paused. This was a set-up question. Everyone gets seasick, and everyone lies when they say they don't.

"No," I answered, maintaining the tradition.

"Would you like to come on a research cruise?"

I paused again before responding, giving my excitement time to release its grip on my vocal chords so they wouldn't squeak. The response eventually came out sounding indifferent: "Sure."

"It would have to be as a volunteer. I couldn't pay you," she added quickly, not realizing that my aloofness was hiding a desire to jump up and down and scream, "Pay? I'd pay YOU for a chance like this."

"That's not a problem. I'll see if it fits into my schedule," I replied, using up any reserve of reserve I had.

"Here's my card. Call me if you're interested and I'll give you the cruise dates." I held it between my fingers. It felt like a ticket to a new life.

3

OVER THE NEXT EIGHT MONTHS, Amy and I spoke on the phone a number of times. At first the calls were strictly about the cruise: what are the dates, what should I bring, and the hundreds of other questions of an excited neophyte. As the frequency of the calls increased, the topics of conversation became more personal. I found out that Amy was seeing someone, a deckhand on another institution's research vessel, and she wasn't that serious about him. This sounded like a thinly veiled invitation for me to ask her out. "Maybe we can have dinner sometime?" I asked one night.

"That would be great," Amy answered, "but I have some things to take care of over the next few days and I have a trip to D.C. next week."

"That's fine. I'll be out of town next week also. We can plan to get together when we're both back." I was optimistically predicting what "things" had to be taken care of.

The following week I phoned Amy in D.C. from my hotel. "I'm flying back into Boston Thursday afternoon," I said.

"So am I," Amy said.

"What time does your flight get in?"

"Two."

"I get in about an hour before. I'll meet you at the gate." I had already decided where we would go for dinner afterward.

"I don't think that would be a good idea." Amy sounded a bit nervous. "Jay is meeting me there to drive me home."

"Things" obviously weren't yet taken care of. I agreed that we could still get together later in the week; but I was never satisfied just being a spectator while events unfolded. At 1:45 on Thursday I positioned myself in the waiting area of gate 23. There were probably a hundred other people milling about, doing what people do to minimize the discomfort of airport accommodations. I scanned the faces, looking for clues that would identify a man I had never met. After passing over the women, I ruled out anyone who looked over 50 or who dressed in a tie. I did not have much experience on ships, but I didn't think many deckhands wore ties to pick people up at airports. The field was narrowed by almost half, leaving about two dozen candidates. Pierced nose—nope. Lavender shirt—not likely. Arms around another woman—unless I totally underestimated Amy's sense of adventure, no. The list grew shorter. Finally I had to use my gut. It told me it was the young guy with his face down in a truck magazine, wearing jeans and a red faded tee shirt.

I should have been content to wait and see if I was right, but I was feeling cocky. I walked up to him and said to the top of his head, "Are you waiting for Amy?"

He looked confused, which was understandable. "Yes," he said hesitantly.

"You must be Jay. I'm David, a friend of Amy's."

His look turned from confusion to suspicion, but before we could go further with the introductions the jetway door opened and Amy stepped into the waiting area. Her vision obviously wasn't that bad, based on the expression of horror she was trying to hide. We continued with some brief introductions and I walked with them to baggage claim, where we said goodbye. I'm sure Jay and Amy had a long ride back to Cape Cod, but I felt I had raised the bar in this budding relationship and it was Amy's turn to take a position.

Amy and I spoke the next day. She didn't hate me. She didn't exactly see the humor in my methods, but she didn't hate me. That was important. We decided to meet a few days later in Boston for dinner. I picked Amy up at South Station bus terminal and we walked along the Charles River. It was overcast, and there were a few flakes of snow in the air even though it was early spring. As we passed the Community Boating boathouse, I heard a familiar voice call my name. It was Craig, an old friend who was helping to run the sailing program there. We chatted for a while and as we were ready to leave, Craig asked, "Do you want to take a boat out? We're not open yet, but there are a couple of Rhodes on the dock."

I felt the cold air on the back of my neck and watched a few more flakes settle to the ground. I wasn't too excited about the idea of freezing my butt on a fiberglass gunwale, but before I could say "Thanks, but no thanks," Amy spoke up. "That would be great!"

We spent the next two hours sailing between the bridges of the boat basin. The snow came and went and the wind was blustery, but it didn't seem cold anymore. Amy took the helm and I trimmed the sails and kept watch for other boats. It felt a lot like the good old days sailing with Doug. We chatted about family and friends. Soon my inquiries became directed toward her studies. She spent the next hour trying to teach me the physics of the Coriolis effect, the apparent force that helps steer the movement of ocean-sized bodies of water. I only grasped bits of the concept then, and to this day cannot explain it to someone else with any accuracy. In between discussions of scientific principles peppered with fluid mechanics, I learned that the way to a scientist's heart is through her research.

As the date of our first cruise together approached, our relationship became more serious. I had long since moved out of the house with Lisa, but remained living near Boston. The weekend commutes to Amy's place on Cape Cod from mine were becoming more frequent. It was May 25, 1995, and I had returned the day before from a two-week business trip in the Midwest. I unpacked suits, ties, and oxfords and replaced them with raingear, woolen caps, and sea boots. The next day would be the start of a new voyage in my life, both literally and figuratively, and I had no idea how far it would take me. It was a gamble, but one worth making. A few weeks before, I had requested a leave of absence from my boss. I didn't have enough vacation time to take a month off, but was willing to do it without pay. His reaction was, "We can't afford to have you gone that long." I gave my notice the next day. I assume they got along without me just fine.

I left for Woods Hole early in the morning. Though well-equipped for my first sail as a novice oceanographer, I had reservations about the adequacy of my skills for such an endeavor. With my bag in the bed of the pickup the drive to the Cape was mile after mile of contemplation on my potential shortcomings as a research assistant.

My truck rolled into Woods Hole under a spring canopy of green leaves, and I started to think seriously about making a permanent move down here. A 60-mile sandbar, the Cape was made up of debris pushed to the front of an advancing glacier. When the glacier receded, it left an isthmus of wide sandy beaches and low vegetation that contrasted sharply with the rocky coast north of Boston. This combined with a relative tranquility and warm water made it an attractive place to vacation. At that time of year, however, the legions of summer tourist had yet to descend on the village of Woods Hole, and the seaside resort was enticing me with its quiet charm. As my truck rumbled over the tiny Eel Pond bridge, there came the realization that before considering a change of residency I must survive my first research cruise. The weather, at least, was providing a good omen. A bright blue sky framed the aged red brick buildings on either side of the road as I looked for a place to park.

I pulled into a lot near the dock. Setting my temporary parking pass on the dashboard, I switched off the ignition and locked up the truck. The walk to the ship was an obstacle course of pallets, crates, and careening forklifts. Stepping onto the Woods Hole Oceanographic Institution dock, I realized that I was as excited about seeing Amy again as I was about going on this cruise.

Sunlight reflected off the small wavelets in Great Harbor like diamond chips floating on the water. Squinting through the glare, I saw the Martha's Vineyard ferry on its return trip from Vineyard Haven, turning into Woods Hole Passage. The R.V. (Research Vessel) *Oceanus* was tied starboard-side-to along the cement quay, her bow rising above my head, held in place by thick rope tied to a massive iron cleat. The information packet from Amy described the ship as 180 feet long and displacing 500 tons, making her a midsize ship in the national fleet of research vessels. Her hull was royal blue and the remaining superstructure a very pale green, the two chosen from a palette of colors the sea is capable of producing. A single stack festooned with antennas and weather sensors rose from the aft end of the pilothouse two decks up. On one side of the stack was the seal of the Woods Hole Oceanographic Institution—a drawing of the original R.V. *Atlantis*, a sailing schooner that took scientists to sea until 1964.

Amy had told me to meet her on board, so I shifted the heavy bags on my shoulders and walked toward the gangway. Each step felt like that of a 5-year-old approaching a fire truck for the first time. All around me scientists, technicians, and crew scurried about, loading equipment, supplies, and provisions for the three-week cruise. Though not experienced enough to realize it yet, I was witnessing my first truism of oceanography: the last piece of gear will always be frantically loaded aboard five minutes prior to departure, no matter how long the ship has been idle in port.

The sailing board hung by the gangway, a two-by-three-foot whiteboard with informational headings neatly lettered in black vinyl. Next to each heading was the variable information that changes with each cruise, scribbled in with a red dry-erase marker.

R.V. Oceanus
Departs: *Woods Hole for sea*
Date: *May 26, 1995*
Time: *0915*
Liberty Expires: *0800*

My footsteps rattled across the gangway that spanned a four-foot gap between ship and dock. This aluminum link to land sat like a drawbridge across a castle moat. Stepping down onto the deck set off a sudden sense of apprehension in me. Piles of odd-looking gear were everywhere and the environment seemed instantly alien. What had I got myself into? I knew nothing about this stuff. What if I screwed something up? Before I could agonize for too long over my ignorance, I saw Amy and a tall, silver-haired man approaching. If a ship is a castle then the captain is king; I was about to be presented to the royal court.

"Any trouble parking?" Amy asked.

"No, none at all. Though I may need a jump when we get back in. Three weeks is a long time for that old truck to sit around." Somehow I was expecting a more personal reunion than talk of the parking limitations of Woods Hole.

"David, I'd like to introduce you to our captain, Paul Howland." Amy's tone was much more informal than I'd expected. I was anticipating some saluting and heal-clicking, or at the very least some "Yes sir, No sirs." Captain Howland seemed, on the contrary, to exude an air of relaxed informality.

I extended my hand to meet his and he shook it with a firm calloused grip.

"Pleasure to have you aboard," he said with the New England lilt that can only be passed down through many generations of Yankee stock. He turned to Amy. "Why don't we meet on the bridge just after we get under way and we can discuss the changes you want to make to the cruise track."

Strangely, as Captain Howland walked away, with only a handshake and the briefest of communication for my assessment, I was filled with confidence in his ability to command this ship. The best skippers of racing sailboats have the easiest rapport with their crews. With nothing else to go on, I made the assumption that this attitude also applied to research vessels. Since the captain bore ultimate responsibility for my safety, my confidence was well-placed.

"Follow me and I'll show you where your cabin is," Amy said. She led me to a massive watertight door with a big round wheel and six locking rods. It reverberated with a loud metal clang as she spun the wheel to unlock it. Amy moved through the ship with a practiced ease that defied her vision problem.

The first thing I noticed after stepping inside was the smell. Greasy, slightly sweet, metallic. It wasn't offensive, just omnipresent, with no detectable origin. It was this combination of fuel oil, lube oil, and painted metal that blended to form the unique aroma that was the *Oceanus.* By the end of the cruise it would be inhabiting every fiber of my clothing.

We were alone inside. "So let me guess, you're not taking me to OUR cabin?" This to the back of her head as she hurried off in front of me. I had made an assumption that we'd be bunking together. We had been, in effect, sharing her house on the weekends, so I didn't think this assumption was too far off the mark.

Amy turned to reveal straight set lips where a smile had been moments before. "I'd like you to do me a favor and not mention our relationship while we're at sea. There are certain rules of behavior on board. This is considered the workplace and I'm your boss."

This boss-employee thing had all the makings of a kinky fantasy, but given the seriousness in her voice I didn't pursue it. I followed Amy down a set of steep stairs to a lower deck. She did not use her cane, but moved along the confined spaces of the ship without hesitation.

"It seems you've got this space mapped out pretty quickly," I said.

"After I get the lay of the ship it is actually an easier place to move around than anywhere ashore," Amy replied, "because everything is bolted down. Obstacles are always in the same place. Besides, I spent a lot of time sailing on her sister ship when I was at the University of Rhode Island."

A narrow hallway ran the length of the ship, punctuated every 20 feet or so by watertight doors that were latched in the open position. Amy stopped in front of the cabin door furthest aft. "You can drop your stuff in here. I'll introduce you to Jason, your roommate, when we get back up to the lab."

My home for the next three weeks was about eight feet by eight feet. Paneled in imitation wood veneer, it was furnished with two bunks, a writing desk, two lockers, and a small stainless steel drop-down sink containing remnants of dried toothpaste from the last occupant. Jason had already stowed his gear and appropriated the top bunk. That was fine with me. I figured if the ship rolled and I fell out I'd have all the fewer feet to travel before hitting the deck.

Outside the cabin and across the passageway, in a space about the size of a closet, was the head. I saw why it wasn't called a bathroom; there was no bath and it was much too small to be considered a room. At first, the head configuration looked cramped. Looking more closely, I saw the beauty of its efficiency: If the ship healed over just right I would be able to take a shower while sitting on the toilet.

Back on deck I watched as the gangway was pulled away. Various deckhands stood by waiting to cast off mooring lines. *Oceanus* sails with a crew of 12 and can accommodate up to 19 scientists. We had a science party of 15 on this trip and the passageways seemed crowded. Sailing with a full complement of 19 must have been a challenge for everyone aboard. The dock at the Oceanographic Institution was the center of activity in the village of Woods Hole, and by the time the ship was made ready to slip her lines a contingent of employees and family had gathered on the quay to send us off.

One long blast of the ship's whistle followed by three short blasts signaled that *Oceanus* was about to back away from the dock. So began our first of 21 days at sea. In my mind it was my probationary period. If things did not go well, my relationship with Amy could have been in jeopardy. Not that I ever expected we would continue to work together; this cruise was just a one-off adventure for me. But one lesson I'd learned from racing sailboats was that living on a boat was an accurate and accelerated test for compatibility ashore.

Standing at the rail with the rest of the science party, we waved to people whose names I didn't know, but I appreciated their presence all the same, as it added a bit of ceremony to the occasion.

On the bridge wing the captain quietly relayed commands to the mate and the ship negotiated the treacherous currents in Woods Hole Passage. Large ferries and small pleasure boats wound their way through the narrow opening into Vineyard Sound, and the captain directed the crew with the solemnity of a priest saying mass. This seemed fitting because for me the whole experience had taken on a kind of reverence. Watching the age-old rituals of casting off dock lines and sounding embarkation signals made me feel like I was in a church service. Commands were chanted out and repeated in acknowledgement, just as had been done by generations of seamen past.

As much as I would have liked to stand on the fantail and take in the vista of a receding Woods Hole, I was reminded by Dan, one of the other research assistants, that there was work to do in the lab. Dan was in his late twenties, and even though he could not have had many years of experience, I got the sense that, to him, going to sea was as routine as going out for groceries.

Before *Oceanus* reached open water every piece of instrumentation and computing equipment that was brought aboard had to be tied down. All hands were busy with pieces of twine and shock cord, trying to find suitable attachment points along the workbenches. Everyone had their own signature method for securing things. Computer monitors, with their ungainly shape and uneven weight, showed the most variation. Dan threw a token piece of string around the base of his, while the person occupying the adjacent space created an elaborate web of knots that made her monitor look like a butterfly larva cozy in its cocoon. This nest-building shipmate to my right obviously subscribed to the "If you can't tie a knot, tie a lot" theory of rigging.

My technique fell decidedly in the middle. While trying to imagine what would happen to this heavy object when the ship pitched, rolled, and yawed, I also pictured what sort of sea state would precipitate all this pitching, rolling, and yawing.

"So what kind of weather do you think we'll see for this trip?" I asked Amy as she felt her way through a box of shock cords, looking for just the right length. Before Amy answered, Dan's laugh filled the confined metal walls of the lab.

"She's not the best judge of weather," he said.

"What's that supposed to mean?" Amy snapped back at him, though with good humor.

"You obviously haven't heard her nickname," Dan said to me.

"Nickname?" I looked over to Amy.

"Don't listen to him. It was a little joke from the last cruise," she said.

"I seem to recall it happened on more than one cruise," corrected Dan.

"What's this nickname?" I asked Amy again, my curiosity piqued.

"Hurricane Amy," she said into the box of shock cord. "We hit some pretty bad weather on the last two trips."

"Sounds like a good name if you're a boxer." I gave a hard tug on the monitor in front of me and thought of hurricanes. What else hadn't she told me?

Oceanus was the largest ship I had ever been on, and despite all the hurricane talk it was difficult to envision the hundreds of tons of steel under my feet moving much. As it was, I found myself stealing glimpses of Martha's Vineyard through the porthole as we steamed east through Vineyard Sound, and if not for the shifting of this landscape I could not, in all honesty, have said we were moving at all.

With everything tied down below I went out on deck into a midday sun. The ship was heading out of Nantucket Sound, passing between low-lying Monomoy Island to the north and Great Point on Nantucket Island to the south. Though there are more than 10 miles separating these two points, navigable waters in Nantucket Sound are at a premium. At the elbow of Cape Cod, a current trying to move this great terrestrial obstruction of sand deposited sediments while periodic storms nibbled at the shore, regurgitating silt someplace else. These ever-moving, shoaling sands surrounding us had trapped hundreds of ships over the centuries, leaving a vast expanse of blue that covered a gauntlet of rocks, bars, and spits. Ship captains must dodge these hazards before making port or risk joining countless hulls littering the bottom. I have sympathy for the mariner of 200 years ago who relied heavily on luck to guide him on his way. Given all the electronics aboard *Oceanus,* I was fairly confident we wouldn't become another name in the shipwreck record.

Dan asked me to help him assemble the rosette, which was the main piece of gear we would be deploying during the cruise. The rosette is not one instrument, but many individual probes, sensors, and collection containers. It takes its name from its shape: a six-foot-diameter aluminum frame upon which are attached 24 gray plastic cylinders called bottles.

In the center of the frame is a device with a series of small hooks around the top. My job at the moment was to rig a monofilament lanyard from the cap on each cylinder to its corresponding hook in the central pylon. The pylon is an electro-mechanical device which is used to release each hook from a command sent down the wire. The top and bottom caps were attached to one another by a large rubber band running through the main body of the cylinder.

After all of the bottles were secured open, the top of the rosette looked like a spider's web, with monofilament line radiating out in all directions. When the entire package was lowered into the water, signals would be sent down the hoisting cable telling individual hooks to release their lanyards. This would cause the top and bottom caps to snap into place over the open ends of the tube, sealing in a sample of water from the desired depth

Other devices attached to the frame, I was told, measured water properties such as temperature, salinity, and dissolved oxygen. One instrument, a large yellow cylinder with four black disks splayed out at the bottom, looked like the compound eye of a fly. This device bounced sound signals off drifting bits of detritus, and using the same principles of Doppler shift found in police RADAR, measured the speed of the current beneath it.

A chilly, freshening breeze had replaced the fine spring weather that bid us farewell at the dock. Since the sea surface temperature was hovering around 50 degrees, the wind was decidedly cooler blowing over open water. I went inside for the requisite cruise kick-off meeting. In the lab Amy and the other Principle Investigator of this cruise, Bob, gave a briefing on the work to be done, our watch schedule, and the tasks that each of us would be expected to perform.

The primary goal of this cruise was to study the body of water known as the North Atlantic Deep Western Boundary Current (NADWBC). Most people are familiar with the Gulf Stream, a surface current flowing north from the Gulf of Mexico well up into the North Atlantic. Even I, novice oceanographer, had some idea of its importance. What is not well understood by the average person is that the water flowing out of the Gulf must be replaced, or we would end up with a gurgling sound near Tampa as the last of the Gulf water drained away. The water for this replacement is fed, in part, by the North Brazil Current flowing up from the northeastern side of South America. This water in turn must also be replaced, and, well, you get the idea.

The oceans are in constant motion, circulating in and out of basins and around continents. The water in the North Atlantic may take hundreds of years to circulate around back to its point of origin. As these parcels of water move about, their properties, i.e., salinity, dissolved oxygen, and temperature, change. This change, however, happens very slowly. Oceanographers look at these properties as a forensic scientist looks at fingerprints, and by taking careful measurements can tell where the water originated. In some cases, other elements have been introduced into the water from external sources; these can include nuclear fall-out and pollutants. These elements, called tracers, can be used by oceanographers to follow the path of water for thousands of miles.

In Boston harbor, scientists studying the dispersal of treated sewage from a new outfall pipe used caffeine as a tracer. The source of caffeine? Eastern Massachusetts's morning cup of coffee. A great number of people around Boston never realized that their trip to the bathroom made them participants in a grand experiment.

This conveyor belt of water is made up of wind-driven surface currents and deeper, subsurface currents that are set in motion by differences in density. Saltier water is denser than fresher water. Colder water is denser than warmer water. Since the temperature and salinity of the world's oceans is not uniform, there will always be differences in the densities of various parcels of water. When these parcels of water are adjacent to one another, denser water slides under the less dense water. This sinking movement is what is known as deep-water formation. As the denser water displaces the water below, it sets it in motion and begins a density-driven current, or, in oceanographic terms, thermohaline circulation (thermo = temperature, haline = salinity).

In the case of the Northwestern Atlantic, the slow-moving thermohaline currents account for a significant portion of the ocean circulation, and because all of this may be happening in water up to seven miles deep, the effects of this movement are not well understood or easily measured.

It is simple to visualize a world of swirling water moving along as surface currents. It is also easy to imagine how to track these surface currents. Drop a message in a bottle, see who calls you back, and you get a good idea of where the current is moving. What complicates the study of ocean currents is that they meander in three dimensions, flowing not only along the surface, but in layers throughout the great abyss. The North Atlantic Deep Western Boundary Current, often abbreviated further as the

"DWBC," that I had been pressed into service to study, is one of these subsurface currents. It flows from north to south at an average depth of 8000 feet. At one point it crosses under the Gulf Stream, moving in the opposite direction. Our goal on this cruise was to measure the characteristics of this water and determine how much of it was being transported by the current. The amount of water that we are talking about is big, really big. So big in fact that we can't even use normal terms to describe it. Gallons won't work; we would need to add too many zeros. Oceanographers have created a numbering system to handle these big volumes of water: Sverdrups. The Sverdrup, named after one of the many great, dead, Norwegian oceanographers, is defined as one million cubic meters of water passing a given point in one second. That's 265 million gallons every second. To put this in perspective, one Sverdrup would fill the inside of the Empire State Building with water in one second. Stand at Cape Hatteras in North Carolina and look east over what seems like a homogeneous expanse of sea. Undetectable in this casual observation is an oceanic highway of water, the Gulf Stream, transporting up to 80 Sverdrups at speeds approaching five miles an hour.

The Gulf Stream rockets along when compared to the Deep Western Boundary Current. The DWBC's one-quarter-mile-per-hour average pace seems no more than a crawl. What sets all this water in motion around the globe and how the different currents interact with one another is, in part, what we were there to find out.

Amy and Bob's study was part of the continuing contribution to the global road map of currents. With each successive study by scientists around the world, the ocean road map becomes redrawn, first with the sea's equivalent of interstate highways, then with local roads and ultimately with seemingly inconsequential dirt paths. The completed maps, in turn, are used by biologists, fisheries managers, meteorologists, and others, as a component in their own research, one more variable in the model of a complex system of earth processes. I felt privileged to be even a small part of the discovery process.

As the cruise meeting progressed, so did the wave height. *Oceanus* was now in the open ocean with no landmass for thousands of miles to the south to interrupt the movement of a northward-blowing wind. I tried to concentrate on the computer monitor that Bob was using to demonstrate the data acquisition software. His boundless energy increased as he moved along. My strength, on the other hand, quickly ebbed, seemingly flowing into him, and I had trouble focusing my eyes. Here in front of me was a man who obviously loved his job, reveled in being at sea, while I sat, nausea welling up inside me, thinking of ways to get off this boat. It is said that the first stage of seasickness is when you think you're going to die. The second is when you're afraid you won't die.

I wasn't alone at the rail. Leaning over and contemplating my lunch for the second time that day, I was joined periodically by a few other members of the science party. Trying to console me, Amy related a story of her experience as a student on a cruise many years ago. The ship's captain found her laying in a locker, a green, semi-conscious wet lump. "This too shall pass" were his words of wisdom. Intellectually, my sailing experience knew this was true. Once the fluid in my inner ear synchronized with the motion of the boat I'd be fine. At that moment, however, my body found no consolation in any of this, and I tried not to dwell on the contents of my stomach, which were again knocking at the base of my esophagus, seeking a way out.

To take my mind off my own misery, I began to think about the equally queasy people around me, professionals who had dedicated their lives to the study of the ocean and were committed to sailing on her back to do so. They had come out here expecting to be sick and uncomfortable and miserable, yet they did it willingly. Sometimes the years treated them kindly and their bodies became acclimatized to the motion, but more often than not, they endured the ritual sacrifice to King Neptune at the beginning of every turn of bad weather.

"This too shall pass," and so it did. There is nothing so enjoyable as the first meal following a bout of seasickness. In these circumstances a can of SPAM would taste like filet mignon. But there was no SPAM. On this trip I was introduced to the cooking of one of the best stewards employed in the research fleet. Hugh was as crusty as the bread he baked. Hard on the outside, but with a soft center. Hugh would stand behind the stainless steel serving table, white apron over his black-checked pants, looking unreasonably gaunt for a man surrounded by so much good food. He never smiled, at least not in my presence, and his nicotine-stained mustache hung unwaveringly straight over his upper lip. Typical was this exchange with Hugh:

"Smells good Hugh, what's in it?"

"What do you care as long as it tastes good? Now move along!"

Working in a galley the size of a garden shed, Hugh and his assistant created dishes worthy of a fine restaurant. The only difference? Here the selection was somewhat limited and no payment was due at the end of a meal, at least not with cash. Such haute cuisine did come with a price however; Hugh had a certain standard of decorum on his mess deck. Violate his rules (no hats, clean shirts with shirttails in) and you would be whisked away faster than egg whites into meringue. Hugh had done galley work most of his adult life. After a naval career as cook serving hundreds of men on a destroyer, Hugh found the transition to steward on a research ship, preparing meals for only 25 people a day, an easy one to make. Not being limited by the confines of a government procurement officer's job, Hugh became more creative in his choice of menu, and over the years he had developed an arsenal of intriguing dishes that made the monotony of sea living more palatable.

It was in Hugh's mess deck that I learned another truism of oceanography: unless the weather was bad enough to initiate puking my guts out, my caloric intake would far exceed my caloric expenditure. The day before I had worried that my body was withering away. The next day I could foresee more than a few extra pounds by the end of the trip. As Amy and I sat down with plates full of prime rib and fresh bread, my arm waved in a vague motion around the room. "So who pays for all this?"

"The food?" She looked at me quizzically.

"No, everything; the ship, the instruments, your salary. It must cost a fortune."

"I have to write a proposal for whatever research I want to do," she explained. "The two biggest funding agencies are the National Science Foundation and the navy. Though since the cold war turned tepid, navy money is drying up."

Amy went on to tell me that the federal government allocated a pool of money for oceanographic research from within the National Science Foundation (NSF), which amounted to about 200 million dollars out of the total NSF budget of five-and-a-half billion dollars. As I cut my steak, I put that into perspective in my own way. The company I had just left helped to build the Peacekeeper (MX) missile. The total cost for this program was more than 20 billion dollars. In the end, all 50 missiles produced were scrapped. The entire government funding of ALL scientific research in this country was less than the cost of this one mothballed weapon system. When proposals from scientists are submitted to NSF, the agency consults with a volunteer panel of other scientists who review and rank the proposals. They then turn this list over to a program manager at NSF, who goes down the rankings from top to bottom, handing out money. When the money runs out the list gets cut off. Many qualified and important research proposals didn't get funding because other proposals were ranked marginally higher on the list, or better fit within the program manager's budget.

Amy had what is known in academia as a "soft money" position. Though her salary was determined by WHOI, and the institution's name appeared on her paycheck, she was for all practical purposes self-employed. If her proposals did not get funded she would have no salary. There are provisions, it turns out, for the Oceanographic Institution to fill in the gaps if lapses of funding do occur, though the expectation is that it is a solution of last resort. Scientists at universities will typically see nine months of "hard money" while they are teaching, leaving them only a few months a year to find outside sources of revenue to support their research efforts and give them a full salary. The stress of this ever-present need to beg the government and foundations for money has driven many qualified scientists away from a life dedicated to research.

I considered doing my part to help conserve funds by not having seconds, but quickly justified another entree after considering my non-existent salary. Over a second helping of desert I read through a summary of the proposal submitted for this project. We had another day of steaming before reaching the first station, and I couldn't wait to begin the real work.

The following morning I ventured out on deck, my camera ready to capture the power and menace of the storm building up around us. The North Atlantic had become a rolling landscape of gray. As the crest of each watery hill rose above the protection of the surrounding waves, wind sliced at its peak and carried off a plume of cold mist.

For protection against the stinging, cold spray I tucked into an exterior stairwell on the starboard side of the ship. I felt the ship slowly climb the face of an oncoming wave. The bow hung momentarily over the crest until gravity pulled us down, accelerating like a roller coaster, into the trough. My stomach, eyes, and inner ear were finally working together to make sense of this alien motion.

Hove-to, angled just off the waves and wind, barely moving so as to minimize the motion and conserve fuel, the *Oceanus* sat off the Grand Banks, a few hundred miles south of Newfoundland. A low-pressure system was creating the foul weather that prevented us from beginning work. Our mission was to deploy oceanographic instrumentation and collect water samples in an area that ran along the continental slope from Canada to New Jersey. My self-imposed mission was to take a picture capturing the awesome size of these waves in a way that would instill fear in the hearts of the people back home. "Wow, those waves are the size of mountains. You were really out in that?" was the response that I was looking for. In reality, even monstrous waves always seem to appear smaller on film, somehow distorted, in the same way that my license photo makes me look like a heroin addict.

With the sea surface temperature around 38 degrees Fahrenheit, hypothermia would paralyze my body within minutes if I fell overboard. Combined with the distinct possibility that it would be some time before my presence was missed, and the fact that in this sea no one could possibly see a tiny human head on the surface, the likelihood of my survival would be small.

Not wanting to set a record for the world's shortest oceanographic career, I squeezed my back closer to the bulkhead. The cold from the frozen metal stair tread seeped through my pants like thousands of fire ants each taking nips at my butt. All of my attention, however, was directed through the viewfinder. I was a voyeur, detached from the maelstrom a few feet away. I was also the only one on deck. *Oceanus* is a wet ship and the rest of the crew was smart enough to spend this downtime warm and dry below, sheltered from wind and wave.

Surface waves are born traveling in the same direction as the wind that created them. Friction between moving air and water causes ripples, which in turn increase surface friction, creating wavelets. These small waves collide haphazardly and either cancel each others' energy or combine to form larger waves. Waves may travel many hundreds of miles before they dissipate. Since the wind does not blow from the same direction all over the ocean, waves from one wind area may arrive in a place with a very different prevailing wind direction. This leads to the occasional "rogue" wave, an unpredictable orphan in an otherwise organized wave pattern. Rogue waves take vessels by surprise and, if large enough, can be and often are the causes of disaster at sea.

The wave that hit the starboard side of *Oceanus* was one of these wayward creatures, though its small size could only produce discomfort rather than danger. It most likely originated in a low-pressure system near the Azores, 1500 miles to our southeast. For a split second, what sounded like the voice of an angel speaking Portuguese whispered above my head. Looking up, I saw this angelic premonition turn into a curl of green water cresting high over the ship's side. A few hundred gallons of cold water hung in space, frozen in time just long enough for me to wish I was someplace else.

4

THE WAVE THAT HUNG MOTIONLESS ABOVE MY HEAD finally obeyed the laws of physics and crashed down upon me, soaking my skin under multiple layers of clothing. I grabbed a railing with one hand to keep from being thrown sideways and thrust my other hand, holding the camera, into my rain jacket. It was no use. Cold Atlantic water drained from my pant legs, taking with it precious BTUs of body heat and giving my camera a good dousing.

As I entered the main lab, wet footprints recorded my wavering path along the metal deck. Being near the ship's center of gravity, this area was probably least affected by the pitching and rolling motion of the boat, though this didn't obviate the need to hold on to every bench top that I passed. *Oceanus* remained hove-to as the low-pressure system slowly made its way east of our position. A few members of the on-watch science party worked at computers in the lab, making sure that all the analytical equipment was ready to go when the captain gave the OK to start deck operations again. Nobody looked up when I walked by, but I sensed them smiling at what was either my novice enthusiasm or complete lack of common sense. Despite my cold, wet encounter with the North Atlantic, there was a sense of privilege about seeing what only a small fraction of people sharing this planet get to experience.

I continued down the passageway toward my cabin, the sound of my squishing sneakers punctuating each footstep. One of the crew admonished me for going swimming, and pressed his back to the steel wall to keep from getting wet as we passed in the narrow hall. I descended the ladder to the deck below, debating whether to walk down backwards given the bucking motion of the ship. They call them ladders and not stairs on a ship for good reason. Easy to master in calm weather, they are so steep as to turn vertical if the ship pitches far enough, creating a wall of treads reminiscent of a jungle gym. Of all the ways to get injured at sea the most common is probably falling from a ladder. Hence the old adage: One hand for you and one for the ship. With this in mind I turned around and proceeded down, a firm grip on each railing.

After a hot shower and a dry set of clothes I went up to the galley for a snack. I met Amy coming the other way through the passage. Exaggerating the tightness of the space, I pressed her body into the bulkhead as I moved by.

"Hey, I told you none of that," she said, but there was a smile on her face and she did not resist.

I looked up and down the passage. There was nobody around; everyone with a cabin on this deck was either on watch or sleeping. "I'm getting some tea. Do you want any?"

"I was heading to bed, but I'll have a cup with you."

"Where's your roommate?"

"On watch," Amy said.

"So she'll be busy for another few hours at least."

"Don't even think about it, someone may see us," Amy said as she pinched a roll of flesh under my ribs.

Amy and I sat down at chairs and a table that were all bolted to the deck. A lip around the table's edge caught my plate of cookies as the ship took another roll.

"Before we left I got a request from a chemist at the University of Washington," Amy said. "He saw our cruise track and wanted to know if we could take some surface water samples at specific locations. It's just a few, do you want to do it?"

"Sure, why not." My own personal project; not bad for a first cruise, I thought to myself. "What do I have to do?"

"In the lab there is a small plastic pail, a piece of clothesline, a funnel and four glass specimen bottles." She handed me a piece of paper. "These are the approximate station latitude and longitude coordinates. When we get close to one, tie the line to the bucket, drop it over the side, collect some water, dump it out to rinse it clean, and fill it again. When you bring it aboard, fill a collection bottle, seal it and label it with the exact latitude and longitude. That's it."

Not very high tech, but it would be my very first contribution to science. "Sounds simple enough."

"Oh, don't forget to call the bridge for approval before doing any work over the side," Amy reminded me.

"I think this stretches the definition of work," I said, taking another sip of steaming tea.

Back in the lab I studied the chart with our course plotted in pencil across the velum paper. The dark line zigzagged across the Grand Banks, projecting a path that ended off the east coast of New Jersey. I carefully calculated our closest point to the first position that appeared on the paper Amy had given me.

When the time for my sample approached I gathered everything I needed on deck. The sea state had calmed considerably and the air was even beginning to feel warm as *Oceanus* skirted the edge of the Gulf Stream. The air retained the smell of the sea, a remnant of the passing storm that left particles of salt water adrift in the wind.

I keyed the transmit button on the deck squawk box. "Bridge, deck," I said in my most professional oceanographer voice.

"Go ahead deck," came the mate's voice out of the metal speaker.

"I'd like to put a bucket over the side to take a sample."

"Sure thing, go ahead."

I picked up the coil of line, and after tying a passable bowline around the handle, tossed it into the sea.

The force on my arm as the 10-knot forward motion of the ship pulled against the bucket almost dislocated my elbow. The only thing that prevented me from losing the bucket was that the line was wrapped around my left hand. With the cord digging deeper into my flesh I had just enough strength to pull the bucket free of the water.

Dan's voice came down from the deck above like that of a boxing coach from the ring's corner. "You know, if you throw the bucket forward, let it sink a little and pull it up just as ship is passing it, it won't try and take your arm off."

"Thanks," I said, wiping a small streak of blood onto my pants.

Thirty minutes later, despite the trauma, I proudly sealed the now full bottles of water of my first research project and packed them in the shipping box. I never knew what became of the information contained in these bottles, and the scientist in Washington probably never knew my name, but it still felt like an accomplishment.

After the first week at sea things began to settle into a routine. The daunting tasks of learning how to handle scientific gear and remembering the new responsibilities of standing a watch were behind me. But even feeling much more comfortable in this new environment, I still had that little uneasy pang whenever I was asked to do anything of importance. Breaking something now, I thought, could not only jeopardize Amy's research, but would probably ruin any hopes I had of continuing a long-term relationship with her. Oceanographers do not go to sea as often as most people would think. A large field program once every year or two is probably average. So in a career of cruises that you could count on both hands, it would be easy to leave a memorable impression, positive or negative. I was hoping to avoid the latter. In this business, dropping another woman's name during sex would have a higher probability of forgiveness than would dropping an expensive sensor on the hard deck.

"We're ready to deploy out here." Amy's voice sounded thin and far away in my headphones, but she was only out on deck, a few meters on the other side of a metal bulkhead separating her from the comfort of my warm, dry chair in the lab. It was dark, rainy, and cold outside and though I couldn't hear the wind, I found its effects on the motion of the ship apparent in the slow rhythmic roll from side to side. My appreciation of Amy's determination had grown in the last few days. She was one of the two lead scientists on this cruise, but she endured the elements and heavy lifting on deck alongside her research assistants. Her vision problem did not seem to hold her back either. I helped with some reading and chart plotting, but Amy seemed in her element on the ship, even more so than on land.

I was grateful for this because it allowed me to graduate from the grunt work outside, handling lines and collecting samples of icy water, to directing the rosette cast from inside the lab. It wasn't difficult to learn, and Dan was patient in teaching me, going over all the small steps that needed to be taken in order to obtain high-quality data.

"You get that Larry?" I asked into the boom microphone that curled in front of my face.

"Yep, ready when you are." Larry's deep voice reverberated in the headset.

Larry was the boom and winch operator. He sat on the open deck above me at the controls of the winch that lowered the package into the water. I picked up the phone and got permission from the bridge to begin the cast. Anything going over the side, from instrumentation to trash, must first be cleared with the officer of the watch on the bridge.

"OK, let's put her over the side and hold one meter beneath the surface."

"Hold at one meter," responded Larry, and I heard the sound of an electric motor starting somewhere above my head.

I hit a couple of keystrokes and the computer began acquiring data from a number of devices aboard the rosette. Temperature, salinity, oxygen, pressure. Based on what I had learned in the last few days, everything looked normal for surface water. "Down at 60 meters per minute. We're stopping at 4800 meters."

"Down at 60 to 4800," repeated Larry.

The trip to the bottom would take a little over an hour. A graph of multicolored lines representing various measurements of water properties began to grow on the screen in front of me. My only jobs were to verify that everything continued to work properly and keep an eye on the echo sounder as we neared the bottom. Ramming these sensitive probes into the mud would result in a quick demotion back to my spacious, but drafty, office outside.

The big steel watertight door to the lab opened and then closed with a clang. Amy sat down in the seat next to me. Her yellow slicker dripped with a combination of rainwater and sea spray. When she removed her hood, wet tangles of hair fell across her face.

"Everything OK here?" she asked.

"Fine, everything looks normal."

Amy leaned over as if to read the data scrolling across the screen, but I sensed the action was purely mechanical; there was no way she could see an image of that size. Given her silence, I suspected that she didn't want to appear untrusting by asking me to read aloud any of the information. In that respect she'd been a great boss, so far anyway. Keeping up this pretense of a strictly business relationship had been difficult. Maybe it would be easier after 10 years, but at that moment, when the attraction between us was still as tall as the flames of a new fire, it seemed impossible to keep these feelings in check.

"Dan and I will be in the galley. Call us if anything seems amiss," she said, getting up.

After Amy left I was alone again in the lab. I took it as a sign of her confidence that she could leave me unattended at the control station. Despite the solitude, it was far from quiet. Noise engulfed the room. High-pitched computer cooling fans, midrange whooshing from the ventilation system, and a low-frequency rumble from *Oceanus*'s engines made an overture of mechanical music. The libretto of this grand opera came in the form of Larry's voice over the intercom.

"So how long have you and Amy been a couple?"

After getting over my shock, I responded, "What are you talking about?" I tried to sound incredulous, but I was thinking, "It's over now, the fat lady just started singing."

"Don't bullshit me. It's pretty obvious. How long?"

"A couple of months. Do me a favor? Don't spread it around."

"Spread it around? Can't hide much on a small ship, and gossip is all I've got out here. No newspapers, no T.V. Whatever you do the whole ship's gonna know about it. Trust me, the harder you try to keep it hidden the faster it spreads. Besides, you do know that anyone can plug into this conversation from any com station on the ship? The bridge even has a speaker up there."

Great! After that exchange trust was the last thing I felt toward Larry. It turned out the intercom system was a perfect conduit for the spread of information. In the long hour that followed, I got Larry's life story through early adulthood. By the end of the cruise, I started thinking, I could be his biographer. Larry had shattered my image of the reticent sailor. He talked so much the wire to my headset began to feel warm.

By then the rosette was on its way back up. I had to relay "stop" commands to Larry so that I could fire a bottle and capture a water sample at various depths. It made for an odd conversation, and after a while we each continued to talk around the interruptions as if they weren't there.

". . . my daughter fights with me constantly about using my car. She drives it. . . ."

"Slow to 30," I interjected.

"Slow to 30 . . . so much when I'm not home she thinks it's hers.

"Stop."

"All stop. My wife is on my case about this list of house projects that never ends. After a while I look forward to shipping out and getting back to sea."

"Ten minutes ago you were complaining to me that you couldn't wait to get home." I pushed a button that sent a signal down the cable to close a bottle. "Up at 60."

"Up at 60. When you're here you want to be there and when you're there you want to be here. Yeah, it's confusing. Give it time. Do a few more trips and you'll see what I mean," the disembodied voice said.

I didn't tell Larry, but at that moment I was having more fun than I'd had in recent memory, and there was no other place in the world that I would rather have been.

5

IT HAD BEEN A YEAR SINCE I'D STEPPED OFF *Oceanus* after my first research cruise with a certain sense of satisfaction. Nothing broke as a result of any incompetence on my part. There were no emergency "man overboard" stops to pluck me out of the sea, and after the brief bout with mal de mere, I managed to be of some use to the scientific effort. More importantly, three weeks of priestly celibacy hadn't killed me, though toward the end, Larry's prediction was correct and there wasn't anyone aboard who didn't see the connection between Amy and me. All of our attempts to keep the relationship quiet, it turned out, were to appease Amy's adherence to the code of shipboard conduct—no sex on the boat. It may be a rule that was broken from time to time by others, but Amy was not one to break such rules. Contributing to this may have been the fact that the first generation of seagoing women oceanographers had not even retired yet. As odd as it sounds, it was not until the early seventies that women went to sea on a regular basis. The prohibitions were still on the minds of many woman scientists, and I think Amy did not want to do anything that would decelerate the progress of the last two decades.

The cruise may have been Amy's test to see how well I fit into the oceanographic lifestyle, or it may have been my witness to her passion for her career, but in either case, it was the catalyst that would take us to the next stage in our relationship.

The ceremony was held at a house on a hill overlooking Woods Hole Passage, a stone's throw from the dock where *Oceanus* was berthed. On that warm September day, in addition to a gold-banded finger, providence provided me with a live-in oceanography tutor. I had not been involved in any more of Amy's work, but she continued to teach me the fundamentals of how the ocean worked, and I walked the beaches of Cape Cod with a new appreciation of what was happening in the water around us.

Since moving in with Amy I had been able to witness the slow but relentless progression of her vision loss. Amy would come home from the office and comment that she couldn't see well that day. I would answer by saying that it was probably the lighting or the stress of the job, but we both knew what was happening. The only difference between us was that I was much more confident than she that she could continue her research despite her blindness.

As time progressed I had become reader, chauffeur, guide dog, and research assistant, with no regrets. All of these things, and my previous experience at the Carroll Center for the Blind, created a new opportunity for me. Putting these skills together, I formed a business providing adaptive technology to blind and visually impaired people. With Amy's encouragement I was able to survive the financially painful early stages of building a company and find a rewarding endeavor to put my energy into. Amy immediately appreciated the added benefit to her of having live-in support for her now growing need for devices to help her see and read. It was a busy time for both of us. There was even the occasional talk of adding "parent" to both of our job descriptions, but there seemed to be no time in this new whirlwind of a life to consider it seriously.

The now familiar path to Amy's office took me past a floor-to-ceiling map of the world's oceans. I could not walk by without pausing to study the midnight blue of the deep trenches and the depth contours that gradually lightened to the pale turquoise of the continental shelf. Fine lines traversed each ocean; some were straight for thousands of miles while others jogged back and forth over short distances. Every line represented the track of a research ship and the location of the historical data used to create the map. Coastal areas were a maze of tangled tracks, while most of the Southern Ocean was unblemished, with only a rare, isolated line visible. These lines were a reminder of how much of the world's oceans had yet to be sampled. Since the map's image really represented a three-dimensional space, the volume of the unexplored ocean frontier, in all its depth, was vast.

I recognized more and more faces of Amy's associates with each visit to the building she occupied when not at sea. Some of them even knew me by name, and I began to increasingly feel part of the community.

One particular day the blue door to Amy's office was ajar, and from inside the room a voice recited a rapid string of words, some of which resembled English. The voice had the pitch and timber of a man's, but it was not a man speaking. It was not a woman either, but an electronic representation that some programmer had attempted to make sound human. Amy eased the strain of her worsening sight by using technology that allowed all of the text on her computer to be read aloud.

I looked in without knocking. She sat with her back to me, eyes fixed on a mammoth, over-sized computer monitor. The height and width of the screen yielded a large image by most people's standards, but Amy was using a software program that enlarged the picture even further. It was like looking at a photograph using a magnifying glass as letters and numbers came in and out of view.

I removed a small, balled-up piece of paper from my jacket pocket and tossed it at the back of Amy's head. She jumped about six inches off her seat, proof of how engrossed she was in her work.

"Don't ever do that again! It's not fair, I can't see it coming," she said, keeping her voice restrained, in consideration of those working in surrounding offices.

"If you could see it coming what'd be the point?" In my opinion this had less to do with being blind than about being able to take a harmless prank. Amy sat there waiting for an apology, something that I was not very good at, especially when I felt like I hadn't done anything wrong. "OK, OK. I'll try to remember not to do it again."

"I won't make you promise because I know it's pointless." Her tone changed. "You may be interested in this. I'm plotting some of the data that we collected on the cruise last spring. Take a look." She turned back to the computer's screen.

Amy had transformed millions of bits of disjointed information into graphical images of how the Deep Western Boundary Current was interacting with the Gulf Stream during our cruise. All of the seemingly insignificant data points of temperature at a given depth suddenly morphed into three-dimensional, multicolored illustrations that showed how the water was moving through the Western North Atlantic last spring. Using these graphical images, Amy would write a detailed description of what she had found and submit it to a scientific journal for publication.

The thought that all of those plots would come together in the form of a scientific article made me feel good. It was proof that in some small way my contribution helped. Proof, too, that I hadn't done anything to screw things up.

Going to sea and collecting data is a small part of what it means to be an oceanographer. For every day at sea there may be a month or more on shore processing and analyzing the data that was collected. It is tedious and exacting work, but if not done properly it negates all the money and effort spent in the field.

Once all of the data are analyzed, the next step is to publish the results. This is the fruit of the scientific labor tree, the point at which everything the investigator learns is synthesized and disseminated to her peers. In the end a scientist may experience a "eureka" moment, that rare phenomenon which fundamentally changes our understanding of life on earth. More often though, the data are collated, analyzed, plotted, and massaged for no apparent purpose other than to publish the results. In areas of basic research where there is no immediate application for this new knowledge, the published papers become threads of understanding that other scientists can use for the basis of new experiments, and to construct more theories. These separate little pieces of science don't look like much, but neither do individual LEGO blocks, small and rectangular with tiny nubs and holes. It is hard to look at one piece and imagine any complex shape, but visit the LEGO museum in Norway, where these blocks have been joined to form wondrous sculptures, and you begin to get a sense of how small discoveries are assembled into useful and elegant theories to describe our world.

Just because the basic building-block discoveries of science are small does not mean that they slip into the academic archives without controversy. As much as the study of oceanography relies on factual and statistical results, the methods for gathering the data and proclaiming a conclusion are still viewed with healthy skepticism within the scientific community. Whether the newly discovered six-finned three-eyed zooplankter is really another species or just a result of all the heavy metals dumped in the Hudson River is a question for the process of peer review.

On the surface, peer review is self-explanatory; a scientist will write up results and conclusions of a study, and his peers will review the information for accuracy prior to publication. Areas of study have become so specialized that the pool of available and knowledgeable scientists qualified to review an article may be small. This leads to a potentially incestuous situation. Just as with a 16th-century European king trying to find a princess for his prince, compromises may have to be made and marriage of first cousins will have to do. For peer review the problem of a limited number of reviewers means fewer checks and balances. It is not unheard of for a reviewer to accept conclusions because they bolster his own research effort. Or he may be overly critical, vindictive of the success of competing work. This is not widespread by any means, and for the most part the system turns out thoroughly vetted results in great quantities. Even with these limitations the process of peer review works. It is better than the alternatives, which would be no review or a review done by some federal agency hiding a political agenda.

After peer review, the presentation of results typically takes two forms: written publication within a professional journal, or an oral summary in front of other scientists. I soon found out that the latter method could be very entertaining to the nonprofessional spectator.

6

EVERY SYMPOSIUM NEEDS A VENUE. Amy and I stepped off the plane in Honolulu, Hawaii, one day before the start of the winter meeting of the American Society of Limnology and Oceanography. Limnology is the study of lakes and not, regrettably, the study of a tart green fruit that garnishes margaritas. This trip was one I was making in an official capacity. Amy could no longer attend conferences without the assistance of a sighted guide. When we began to look into her needs for someone to help, we realized that it made the most sense for me to go. I was by then familiar enough with her research, the terminology, and the people she worked with to be able to guide her through the chaos that these conferences generated. As an added bonus, I could set up and keep her adaptive equipment working. Despite the justification for my being there, standing outside the airport terminal building, with a radiant sun on my face, I felt the need to look up "boondoggle" in a dictionary. It was during this moment of sun worship that two of Amy's colleagues from the University of Rhode Island walked up to us.

"Nice to be in warm weather for a change," said one.

"I'm ready for winter to end back home," added his companion.

More oceanographers joined us, all easily identified by the three-foot-long cardboard tubes containing posters of work they would be presenting later in the week. A large percentage of the swelling group was made up of scientists from Woods Hole.

"It was starting to snow when we left, picked a good time to get out of New England," said a fresh face in the crowd.

Murmurs of agreement percolated through the now burgeoning throng of marine biologists, chemists, geologists, and physicists. Looking at the faces around me, I began to realize that there were more years of higher education standing at this bus stop than there were in the entirety of some undeveloped countries. Two years ago, my limited association with oceanographers would have made this encounter intimidating, but after getting to know Amy and her colleagues I found that I could hold my own in social situations. I just kept my fingers crossed that dinner conversations didn't stray too deep into fluid dynamics.

Who picked Hawaii as the location for this event? As more and more of Amy's fellow scientists arrived from places like England, Germany, Miami, Washington, and California, it became apparent that this site was only convenient for the few oceanographers who resided in Hawaii, and the one guy from Tonga who studied the reproductive cycle of the conch. It was here that I learned another oceanographic paradigm: The dissemination or collection of scientific information is important even when done in a place like Iowa, but if it can be done in a beautiful setting near the beach, all the better.

The meeting was being held in the Honolulu Convention Center, a huge complex of glass, concrete, and palm-lined atriums. Lava-stone waterfalls cascaded from the high ceilings, reminding conventioneers that there really was a paradise outside if they took the time to go out and look. The building was more than large enough for the 2200 conference attendees. Scanning the scene, I was hard-pressed to find 2200 people. Maybe some of them had yielded to the call of paradise, and were snorkeling at Diamond Head.

The conference sessions were divided into different areas of scientific interest. There were approximately 12 sessions per day, all running simultaneously. Each session comprised a dozen or so

15-minute talks. The scientist, with the aid of an overhead projector, laptop computer, or other multimedia device, was expected to distill a year's worth of blood, sweat, and sea water into a quarter of an hour's worth of results. Subtract 30 seconds for fumbling with the microphone, 30 seconds to figure out why a slide is upside down and backwards, and two minutes for questions at the end, and the speaker is left with a scant 12 minutes. In these 12 minutes, roughly 30 seconds for every month of effort he probably spent working on his project, his peers would pass that elusive and subjective judgment of his worth as a researcher.

Amy needed to review the day's sessions and I began reading the titles in the list of abstracts.

"Biochemical linkages between rapidly urbanizing coastal watersheds and the coastal oceans."

"Next," Amy responded.

"Mediation of benthic-pelagic coupling by life-cycle patterns and vertical mixing."

"Skip over anything with the words 'bio,' 'pelagic,' or 'coastal,'" Amy said.

A quick skim of the program book revealed "Quantification and regionalization of benthic flux rates: Implications for ocean budgets."

Amy sat up a little straighter, her interest piqued. I was still trying to figure out how I'd said it all with a straight face.

"Let's read some of the talks in that session," she said.

"Spatial and temporal variability of benthic fluxes in the North-Western Black Sea," I read slowly.

To label some of this research as esoteric would be an exercise in understatement. That is not to say that the work was unimportant. It is only from these small, sometimes incomprehensible papers that scientists can assemble pieces of the big puzzle. For me, with little knowledge beyond basic concepts, the prospect of sitting through days of such material was daunting. I challenged myself to try and learn whatever I could while minimizing any audible snoring once the lights went down.

Some of the talks *were* interesting, of course. Global warming and climate change studies were big topics, and from many speakers valuable information on the state of the global warming debate could be deciphered. The remaining discussions provided me with a significant amount of entertainment, if not scientific knowledge. While all of the attendees possessed high academic credentials, the presentation skills of many were subpar, and some of them would probably have had a hard time explaining gravity to Isaac Newton.

We sat up front in the first row of the auditorium so that Amy could attempt to read the material presented with the aid of her monocular. The 10-power single-barrel lens did bring things closer, but at the expense of a narrow field of view and reduced light transmittance. Overall, it had limited benefit. My job therefore was to create a verbal picture of what was being projected and whisper it in Amy's ear during each presentation. The success of this method was directly related to how well the speaker presented his or her material, and how fast I could interpret what was on the screen.

When lots of obscure equations and convoluted charts were displayed I found myself leaning over to Amy and saying, "The train just left the station, and I'm not on it." At other times when it was obvious the graphics were a real mess I told her, "Just be glad you can't see this."

The first speaker of the most recent session we entered advanced to the podium carrying a two-inch-thick ream of transparencies. After spending 10 minutes on his introductory slide, he seemed genuinely surprised to be told that he had only 2 minutes to show the remaining 30. His solution: whisk every slide across the projector at a rate of one every 1.5 seconds. The resulting blur of information was matched only by the rate of his speech, which was by then unintelligible.

The second talk was by a young graduate student wearing a pair of tight pants and a shirt cut high enough to expose her pierced navel. She was obviously nervous, though after seeing the first presenter she should have felt confident that she could not do any worse.

Her presentation was on a laptop computer, displayed on the large room screen via a video projection device. I found my eyes distracted by the glint of the projector's beam off the gold ring in her belly button, so it would not have been fair for me to try to judge the content of her material. In any case, she spent the balance of the presentation with her back toward us, talking at her keyboard while bending over the laptop. Her butt swayed in our faces not three feet away. My first reaction was to describe the scene to Amy by having her visualize a stripper performing atop piano bar, but Amy had probably never been inside a strip club and would have a hard time conjuring up such an image. I settled for telling her that things were not looking very professional.

"Even I can see that," Amy whispered back in my ear, then added, "If you're expecting more displays like this I'm afraid you're going to be disappointed. I only hope, for this woman's sake, that her adviser isn't in the audience."

After the woman finished, I didn't know whether to applaud or tuck a folded dollar bill into the waistband of her pants.

More bad lectures followed, punctuated by the random insightful, concise, and well-prepared ones. One example of effective communication of scientific information was Amy's presentation. I will freely admit to being biased here, but having watched Amy put this presentation together in the weeks before the conference and seeing her spend days distilling pages of notes, tables, and diagrams into a 15-minute summary of only the pertinent information, I knew that she deserved the compliment. After all of this thorough preparation, she had sat me down with a watch and made me time her rehearsal. The goal was not to script every word she would recite, but to instill in her own mind what points to make for each slide she showed. In the hotel room, the morning of her presentation, I could hear eloquent descriptions of float trajectories emanating from the bathroom, and I learned a great deal about the concepts of vorticity, turbulent mixing, and isopycnal ballasting over eggs at breakfast. This last recital went a few minutes too long and meant some relentless last-minute trimming with digital scissors to pare down her presentation into something approaching 12 minutes.

It was time for Amy's presentation and I was seated in the front row of the auditorium, suffering from sympathetic pre-performance butterflies. Always in awe of Amy's presentations, I knew what most of the people in the room did not: Amy has to memorize everything. Her vision problem precluded her from using any visual clues on the screen. She couldn't read notes and couldn't even see the details of her own figures projected 10 feet tall over her head. The only option for Amy was to commit to memory the text of every slide, and have a mental copy of each graphic to the extent that, if questioned, she could point to the correct location on the screen.

Amy stood in front of an audience of about 75 people. She was presenting a condensed version of a recently published paper on the movement of water from the Mediterranean Sea into the Atlantic Ocean. I felt a great sense of pride watching her as she began to talk. It is one thing to submit results in print and maintain some distance from potential criticism, but quite another to stand in front of a live audience and be subjected to critique in real time. Rare as it was, I had seen some presenters reduced to a stammering mass as they withstood a barrage of hostile questions.

The room light was subdued but it was not totally dark. Amy's hair showed in silhouette against the bright backdrop of the projection screen. She began in a confident voice which gave no indication of nervousness; I must have swallowed the butterflies for both of us. She moved smoothly through her introduction to a description of the path of water flowing past Gibraltar and around the Iberian Peninsula. She swung her arms in great circles to accentuate how this water formed large swirling eddies.

These eddies, nicknamed meddies for MEDiterranean edDIES, average 100 kilometers in diameter and one kilometer thick. They form off the southwestern corner of Portugal at a depth of about

1000 meters and continue to spin their way into the Atlantic for several years. When the energy within the swirling mass diminishes, these clumps of warm, salty Mediterranean water stop turning and finally mix completely with the colder, fresher North Atlantic water. The importance of this movement was not yet known. The contribution of saltier water to the Atlantis could influence fisheries, climate, pollution transport, or many other things. At the very least, this phenomenon becomes another piece of the model we use to understand our planet.

Amy continued, unpressured by the clock on the wall she couldn't see. Her laptop sat at a table to her right, connected to the video projector that illuminated her face like a spotlight whenever she moved into the beam. Her expression showed that she was having fun. Amy almost never turned around to face the screen as she changed slides with a click of the mouse. Projecting out to the audience, she described the important points of various figures, all the while conveying the sense of excitement she felt for her work. The slide titled "Conclusions" flashed on the screen and I took my first breath in 15 minutes. My confidence in Amy's ability did not negate my right to spousal anxiety.

In the end the audience provided subdued applause. No rousing ovations here. The results, while interesting, were not about to win Amy a Nobel Prize. I turned around to see how many people were walking out, and was happy to count a fair number of bodies heading for the door. Those people had most likely stayed specifically for Amy's talk. If large groups of people leave after a speaker finishes, it can be quite a blow to the ego of the next person presenting.

Amy sat down next to me.

"Excellent, smooth as silk," I whispered in her ear.

She let out a sigh. "I forgot to mention. . . ."

The remainder of her sentence was lost in the introduction of the next speaker, but it had to do with a quadratic equation or something like that. I reminded her that it was a good thing to leave them guessing; otherwise there'd be no incentive to buy the book. She gave me that "don't joke about my career" look as the next speaker spilled his pile of slides on the floor. We finished out the remainder of the session and grabbed a quick lunch before heading to that other ubiquitous element of scientific conferences—the poster session.

The convention hall's floor was covered by a vast expanse of blue carpet, a fitting color, for it looked like a sea of oceanographers before us in the mammoth room. Hundreds of people milled about, walking through a labyrinth of seven-foot-high presentation boards. Each board was connected to another, forming a chain that gave the impression of a garden maze on an old French estate. Navigation of the maze was made only slightly easier by tall poles indicating the section numbers of the posters in the group below. There was, I found, a method to the puzzle. Walking among the felt-covered boards, I could see that the posters were grouped into sections of similar research disciplines, just as the oral presentation had been.

Posters are just what the name implies: large banners with text and graphics representing the just-published or soon-to-be-published work of the scientist presenting the poster. Over the years, the poster has evolved from hand-drawn paper sheets or copies of pages from a published journal to wall-size, computer generated, 16-million-color eye-candy. The scientist can use the board space as she wishes, posting her results and conclusions for people to read. Typically the author will stand by his poster for much of the afternoon, answering questions and entering into a dialog with other scientists who show an interest in the work.

Some scientists had large groups of people crowded around their board, while others looked like wallflowers at a dance. Some chose to fill their entire allotted four-by-eight-foot space with 10-point text, while others showed 300 graphs that looked so much alike my eyes began to cross trying to find a difference.

Amy and I walked the endless rows of posters. Experience had taught me to identify what poster subjects were of interest to her. I cringed when an interesting subject was presented as pages and pages of text: I would have to read it all out loud, and I was beginning to tire of the sound of my own voice. Because of all this reading, I found myself becoming critical of each poster we stopped to view, and was starting to feel like the judge in a high school science fair. The fact that my brain comprehended only about 10 percent of what was presented didn't deter me from being judgmental. I felt I had earned the right.

Amy became engrossed in conversation with a German scientist who shared her interest in sub-polar currents. She was familiar with his work and had reviewed the information he was presenting, so my reading services were not required. I took the opportunity to wander around alone.

A few rows over, a poster caught my eye. Enlarged photos of intricate, multicolored crystals were tacked up over pages of text liberally seeded with long chemical symbols. The photos were fascinating. There was an artistic quality to the complexity of form and symmetry of shape in each specimen. Golden hues of amber, rose, and sage radiated out from the backlit images.

"Are you interested in diatoms?" An older man approached from behind me. He looked at me through deeply crinkled eyes, probably the result of years of looking through a microscope, or of being someone who laughed at everything.

"I don't understand enough about them to know if I'd be interested," I replied.

"Not your field?"

"I don't have a field," I said, not wanting him to waste his time with someone only window-shopping for science trivia.

He didn't seem put off by my admission of ignorance. Whether it was the obvious lack of a crowd in front of his poster or an excuse for him to preach his message, he seemed happy to share his love for diatoms with me.

Diatoms are single-celled algae. They inhabit every body of water, fresh and salt. The images on the board represented fossilized remains.

"So tell me, why should I be interested in diatoms?" I asked.

His face turned serious, the same reaction my math teacher had when I told him I could never imagine a situation in my future in which I would need algebra.

"Well," he began, "like other phytoplankton, diatoms use sunlight to convert carbon dioxide and water into food. Between diatoms and all the other phytoplankton floating around, the photosynthesis that keeps them alive converts as much atmospheric carbon to organic carbon as do all the plants on land." He stopped, waiting for me to jump up and down with excitement. All I could muster was a blank stare.

He continued, trying to withhold his exasperation at my apathy toward his life's work. "In the mid-19th century people only appreciated diatoms for how they looked under a microscope. Now we understand that they may have big role in the control of greenhouse gasses—in this case carbon dioxide."

"Diatoms may slow global warming?" I asked with hesitation.

He thrust his hands in the air, happy that he'd finally connected the synapses in my brain. "The oceans absorb a huge amount of carbon dioxide. Diatoms are large. The largest ones can be half a millimeter in diameter. They're huge compared with other phytoplankton. Since there are a mind-boggling number of these buggers out there and they don't live very long, they take millions of tons of carbon dioxide with them when they die and sink toward the sea floor. We would be in a much bigger global warming mess today if it wasn't for diatoms."

"Thanks, you've got me interested," I said to him. I shook his hand to leave, but not before taking a closer look at the photographs with a new appreciation.

"Always happy to make a new convert," he said as I walked away.

Retracing my steps back to Amy, I started to think that maybe the world would be a better place if we had less religious, and more scientific preaching like what I'd just experienced. If scientists could capture the zeal of evangelicals, maybe they would feel empowered enough to lift a few politicians up by the collars and get them to take notice of what was happening in the world.

I got the sense from this meeting that posters were deemed by many people in the profession to have a lower status than oral presentations. Even the word "poster" sounded inferior, like something you would find plastered on the wall in a music store. Scientists will often say: "Yes, I'm going to the conference, but I'm only giving a poster." What they really wanted to say was: "If the organizers knew anything about the importance of my work they would have begged me to give an oral presentation."

This seemed too bad, because the less formal environment of these poster sessions appeared to be a more effective way for scientists to convey the details of their results to one another. By only talking to a few people at once, a scientist has more time for detailed questions and answers, and a real dialogue can develop. This lack of structure and time limits allows for a more free exchange of ideas, which was after all the whole point of the conference.

By the latter part of the week, I was beginning to understand that the quality of the presentations was less important than the fact that so many like-minded scientists were all there together. For a profession that was characterized by a great number of solitary months hunched over a computer in a dark office, this conference was an opportunity for people to get together and share ideas—not just in front of the auditorium, but at lunch, over dinner, and maybe even while snorkeling at the beach.

7

I WAS PACKING MY BAGS IN GIDDY ANTICIPATION of a second research cruise. It had been four years since Amy had been to sea on a major expedition. She had spent most of the time since our wedding analyzing and publishing data from previous cruises. My own business had become stable enough for me to place it in the hands of someone else for short time, giving me the opportunity to go with Amy on this trip. I was enjoying what I did, but the thought of going to sea again was worth any financial hit to my bank account. My role as Amy's guide was becoming increasingly necessary. Her vision had continued to deteriorate. The decreases were slow and incremental, a fact that was sometimes a source of great frustration. Amy announced on more than one occasion that if total blindness was in her future, then she would rather it happen suddenly so she could make the necessary adaptations and be done with it. That attitude wouldn't last long, and she would always come to the conclusion that it was better to hold on to every last functional retinal cell with all the strength she could muster.

We were heading to Africa, near the equator and far from the cold North Atlantic waters of the previous trip. Knowing this, I eliminated all the heavy, cold-weather portion of my wardrobe. I sneered at my thick insulated rain gear as my hand brushed past it to locate a lightweight tropical set. I finally decided to take neither, surmising that if it rained at all it would probably be a respite from the equatorial heat. Better to be wet and cool if the opportunity presented itself.

Travel of any length requires some amount of advance planning. For this trip, our first of two cruises to the Gulf of Aden, Amy had spent the better part of two years making preparations. Her whole life revolved around planning, which luckily was something she was very good at, though it sometimes grated against my more spontaneous tendencies. She planed what time to get up in the morning, how long it would take to eat breakfast, and how she would get herself to work that day. There was a sub-plan to the master plan that involved when she would retire and how many children she would have. Somewhere in all that planning was the plan for the Red Sea Outflow Experiment, one that required the transport of tons of equipment and a dozen people halfway around the world. On the kitchen table I picked up a long list of items printed in big letters in bold, black ink—Amy's writing. The Apollo astronauts didn't have as long a checklist to get their spacecraft to the moon as we had for getting our gear and ourselves to the Big Continent.

Later that evening Amy announced some troubling news. "I've got a small problem," she said.

"How small?" I asked, taking this as my cue to pour her another glass of wine.

"We've got clearance to do research in the waters off the coasts of Yemen, Djibouti, and Eritrea, but Somalia is a problem. There is no government there to ask for clearance."

"What do the Ship Operations people say?"

"They're talking to the State Department now. If they say no, the cruise doesn't happen and I don't know what I'll do. I've spent two years trying to get there." Amy looked pensively into her wine glass.

By international convention every coastal country is given a 200-nautical-mile limit from its shores that is defined as an exclusive economic zone (EEZ). If a 200-mile gap does not exist between two countries, then a line is drawn down the middle of whatever body of water separates them. Scientists must respect a country's EEZ and petition that country's government to do any work within it.

We were two days from departure. For Amy it was two days of phone calls, faxes, and meetings trying to resolve the issue. The cruise would not actually be in Somali water for another few weeks. There was time, but not much. It was not unusual for research clearances to be granted at the last minute, even if the request was made months in advance. Sometimes the delay was caused by bureaucracy; other times it was a country flexing its diplomatic muscle just because it could. In the case of Somalia, there was no entity there to petition, which left the U.S. government with a decision to make: go without clearance, or cancel.

In the end, the State Department and the Woods Hole Oceanographic Institution come to an agreement that allowed the research vessel *Knorr* to operate in Somali waters, despite the anarchy and chaos boiling within the country. Arrangements were made to have a private company provide additional security for the ship, just in case. In case of what, I thought? Would a country without a government care enough about a clearance to molest us for taking a few liters of water off their coast? To me it all seemed like a paper-chasing exercise, and I was too excited about the rest of the trip to give it any more thought.

When traveling for a week, or two, or even three, Amy and I pretty much threw some extra fish food in the tank and locked the house. Passing the magical 40-day point required that we delve deep into our collective subconscious to identify all of the daily activities we did by rote and plan how to handle them while we're away. Forgetting to cancel the newspaper when gone for a week results in seven piles of soggy yellow newsprint, looking like droppings from a paper maché deer. Go for a month and we're cleaning up after a paper maché elephant.

On the morning of our departure Amy took the dog out for a last walk. His name was Bingo, a title he inherited at the pound and one that fit him well, since he was more suited to the farm of the children's song than to any real farm. There remained only a few scattered articles left to pack in my sea bag. What's done is done, I thought to myself. The ball was in motion and if it rolled out of our control, so be it.

Loud hammering from the room above reminded me that we were also leaving a significant mess behind us. Sawdust had settled on everything we owned. The contractor assured me that in the eight weeks we were to be gone all the renovations would be completed and we could look forward to returning to a clean new house. Yep, things were moving like clockwork. In another hour my parents should arrive to take us to the airport and take our dog for a two-month vacation in Maine.

Just as I was feeling that this was the least stressful beginning to any trip I'd taken, the front door burst open, leaving a dent from the knob in the plaster wall. Amy's voice resonated through the house with the tenor of a woman who had been physically attacked.

Rushing into the living room, I saw my wife framed in the doorway holding a straining leash that was connected to our cowering dog. His black and white coat was flecked with the foaming saliva that hung from his jowls.

"He got sprayed by a skunk!" Amy shrieked, her voice cracking. Tears rolled down her cheeks. Whether this was caused by her state of mind or the acrid vapors rising from the dog, I could not tell.

No sooner were the words out of her mouth than the toxic cloud enveloped me. "Calm down, calm down. We can take care of this," I said, breathing through my mouth. This has always been one of the finer characteristics of our relationship; one of us always manages to remain rational whenever the other is ready to jump out of his or her skin.

"He got it full in the face. I thought he was sniffing at a cat, and by the time I could see what it was he yanked the leash out of my hand," she shouted with a mix of anger and concern. "He's been coughing and wheezing all the way back. Is he all right?"

Bingo's ears hung by the side of his head as if weights were clipped to the tips. His tail was gone, tucked somewhere between his legs. Mournful, mucus-filled eyes looked up at me. Drool dangled from his mouth and his nose was dripping, but other than that, he seemed OK.

Gagging, I took the leash from Amy and dragged the reluctant dog toward the bathroom. "Let's get him in the tub," I said, not knowing where else to start. From upstairs came the sound of windows being opened as the carpenters, now choking on the fumes, rushed to find fresh air. "Close the bathroom door behind me," I told Amy, not wanting to contaminate the whole house. "I'm going to throw on a pair of coveralls. Don't let him out of the tub."

I armed myself with as many different soaps as I could find. When I got back to the bathroom I knelt by the tub. "We don't have any tomato juice. Go see if the neighbors have any," I said to Amy. With a pair of bright yellow rubber gloves on each hand I started lathering up a trembling bundle of fur.

A few moments later Amy returned and handed me two large cans. The labels read TOMATO SAUCE WITH ITALIAN SEASONING. I looked up at her. "They didn't have any juice," she said.

As the contents of the first can poured onto the dog, big pieces of tomato fell on his back. Turning the can around, I saw the words CHUNKY STYLE appear in small letters. At this point either the weight of the chopped vegetables or the smell of oregano ignited the dry-off response in the dog; he began a shaking motion which originated in his nose and propagated to his tail.

Bright red sauce and chunks of tomato flew in all directions. Some landed on me, but most of it adhered to the white walls of the shower. I looked at the red streaks sliding down the tile and felt as if I was in an Alfred Hitchcock movie. This is one more thing I had to clean up or someone would think we'd committed murder and fled the country.

With Bingo rinsed off I leaned over him to take a deep whiff. We'd gotten somewhat used to the stink while we were confined in the bathroom, but the skunk smell was still prominent, only now he smelled like a lasagna that got sprayed by a skunk.

Leaving the bathroom, I saw the backs of three men hurrying

out the front door. "I'm sorry, but we can't work until the place is aired out," the lead carpenter said without slowing down. As I searched for a phone directory I began to wonder how much this supposedly "free" dog from the pound was costing me in idle labor. After a few frantic calls I found a groomer sympathetic enough of our plight to take him on short notice. By the time we arrived back at the house my parents were there. The dog was relatively sweet smelling and sat there happily, proudly showing off the blue bow around his neck, a gift from the groomer.

Getting into the car, it was my mother who made the observation that while the dog didn't smell that bad, Amy and I did. As the car pulled out of the drive I glanced at my watch. No time to change clothes. It was going to be a very long flight for the people who had to sit near us on the trip to Africa.

8

THE RED SEA OUTFLOW EXPERIMENT, or to use the Boston-biased acronym, REDSOX, was a program designed by Amy and her colleagues at Woods Hole and the University of Miami. Costing $2,000,000 and spanning four years, REDSOX was created to help quantify how warm saline water from the Red Sea is transported and mixed with water in the Gulf of Aden. This heat and salt ultimately diffuses into the Indian Ocean with a clearly discernible signature. The impact of this mixing is unknown, but it could be significant, as the process may contribute to global ocean circulation. In addition, by knowing the impact of the outflow, scientists can speculate on the consequences of the outflow being dammed by manmade or natural causes. This scenario is not as farfetched as it may seem. Ocean heights have risen and fallen over the millennia, forming land bridges out of shallow sills such as the one at the southern end of the Red Sea. Loss of the influx of Red Sea water would change the water properties in the Gulf of Aden and could in turn alter the climate in the region. Also, knowledge of the behavior of water flowing between these areas now helps oceanographers understand what life was like millions of years ago. This niche science of looking into the past, called paleo-oceanography, helps us predict what will happen in the future.

This would be the first of two cruises to the Red Sea. The second one was scheduled to take place in the summer for the purpose of determining if there was seasonal variability in the way the water flowed. We would be working in an area that is defined by the Gulf of Aden, though the principle water mass we would be looking for was Red Sea water flowing into the Gulf.

Unlike my *Oceanus* cruise that departed from and returned to Woods Hole, this trip required that we meet the R.V. *Knorr* in Durban, South Africa. Most oceanographic research takes place in parts of the world that are far removed from the research centers of the United States and Western Europe. To increase the efficiency of the research fleet, a vessel may work in one region of the globe for months or even years at a time without retuning to its home port. This means that scientists and crew for most cruises will have to transport themselves to the ship for their tour of duty. This arrangement also necessitates moving great amounts of heavy and fragile scientific equipment around the world.

For this reason, in addition to scientific merit, the funding of research projects takes into consideration the likelihood that they can be scheduled together with other projects in close geographical proximity. The cruise that we were about to embark on was a prime example of the logistical challenges that face scientists looking to work in remote or hostile areas.

Sitting between the Arabian Peninsula and Africa, the Gulf of Aden does not at first glance appear to be an isolated, remote oceanic outpost. A look at a political map, however, brings the region into focus, and the long narrow gulf looms open like the mouth of a wolf prepared to swallow any ship that enters. With the Gulf's borders defined by the beaches of Yemen, Somalia, Eritrea, and Djibouti, the turbulence on land far exceeds any we can imagine finding under water. Piracy in these waters is not uncommon. Drinking the water in India may lead to discomfort, but merely sailing on the water in places like the Gulf of Aden or the Straights of Mallaca in Indonesia has been known to kill a ship's crew. Nonetheless, the Gulf sees a tremendous amount of shipping, as all ships heading to the Suez Canal must pass through it and into the

Red Sea. Due to the propensity for conflict in this area and the terrorist attack on the U.S.S. *Cole* in Yemen a few months before, the Woods Hole Oceanographic Institution (WHOI) determined that the closest port that could be considered even remotely secure was Mombassa, Kenya, 1000 miles away. Since the U.S. Embassy in Kenya was bombed only 18 months prior to our visit, I reflected upon how bad the alternate ports must be.

Knorr is technically owned by the U.S. Navy and is operated by WHOI, using funds supplied mostly by the National Science Foundation (NSF). On paper it looks to be a complicated arrangement, but it works quite well. WHOI takes care of the ship and provides the crew, while NSF, the Navy, NOAA, or any other institution can request ship time whenever they need it and have the money to pay for it. Considering the risks involved in working in these waters, WHOI did not make the decision to send its flagship research vessel into the Gulf easily, and only granted permission when it was decided that a team of security consultants would be brought aboard in Kenya. Two paid security professionals would meet us in Mombassa. Not having experience with paid mercenaries, those of us leaving Woods Hole were taking bets; would these guys look like over-muscled Rambos or vodka-martini-drinking James Bonds?

The preceding expedition of the *Knorr* was to end in Durban. Any equipment that could not be loaded on the ship when she left Woods Hole four months earlier had been sent by air to South Africa. Amy and I planned to oversee the loading of this gear onto *Knorr* in Durban and then visit some of South Africa before flying to Mombasa, where we would rejoin the ship with the rest of the science party.

With nothing left to do but get to South Africa, Amy was feeling palpable relief. Whatever could be done had been done; the rest lay in the hands of the freight company and the ship's agent in Africa.

We settled into our seats on the crowded Boeing 747 that would take us to England. Anticipating a level of discomfort that can only be achieved by 30 hours of plane travel, I found myself bristling at the flight attendant's chipper voice scratching its way through the intercom. There must be some lesson in flight attendant training school that deals solely with the phrase "once again." Upon graduating, a flight attendant cannot begin or end a sentence without using this phrase. "Once again we would like to thank you for flying with us," "Once again please fasten your seat belts," or my favorite, "Once again let me be the first to welcome you aboard." Of all the world's professions, only a flight attendant can "once again" do something for the first time.

There was no easy way to fly to South Africa from Boston. A series of overnight flights stopping in London, going on to Johannesburg, and then ending in Durban was for us the most civilized route. But whatever way we went we would once again be bombarded by a series of once agains, not only in English, but in French, Dutch, and Zulu.

I tried in vain to sleep on the long trans-Atlantic flight, ultimately giving in to the excitement of our upcoming adventure, and instead watched Amy snore away blissfully.

9

As our plane descended over Durban I peered out of the scratched plastic window. Shining buildings rose from the sand along the beach below. I didn't know what I was expecting, but somehow the image in my mind was more "African."

"What do you see?" Amy asked, leaning across my seat in a futile attempt to see for herself.

Feeling cheated out of a fitting introduction to the Dark Continent, I answered, "It could be San Diego. I think we got on the wrong plane."

It felt like the time I kissed a French exchange student in seventh grade. It was good, but not the exotic French fantasy I had envisioned. Durban from the air could be any number of U.S. oceanfront cities. Among the high-rise pillars of glass and concrete lining the beach, one stuck out, a neon badge proclaiming "Holiday Inn."

On the way to the hotel, our cab driver, Pieter, found out that we were here to meet a ship. "Two nights ago, a ship captain—Greek—he was stabbed to death while walking back to his ship."

Amy's hand, which had been resting gently on my leg, clenched my knee tightly. "Does this happen often?" she asked.

Pieter lifted a bony hand from the steering wheel and waved his fingers dismissively. "The captain, he should have known better. Durban is a nice city, but we have a lot of poor people here.

Poor people are desperate people. My advice to you: don't walk the streets at night."

He didn't answer Amy's question, but before we could press him further the taxi pulled up in front of the hotel. Too tired after almost two days of air travel, neither of us was in any condition to explore Durban's night life and we retired to the safety of our relatively plush accommodations, 23 stories above the city's streets.

The next morning a short taxi ride brought Amy and me to the Port of Durban. A security guard at the main gate waved us through. Our driver dodged deep, oily puddles in the rutted asphalt as he searched for the pier where *Knorr* was berthed. Ships in various states of decay lined the quay. Newer vessels were being loaded or unloaded, while older ones sat abandoned, waiting for crews that had long since gone. We entered another open chain-link gate peaked with barbed wire that was mostly rusted away. Two laconic guards barely looked up as our taxi passed them.

Knorr was tied up smartly alongside, her fresh paint contrasting sharply with that of the scows at her bow and stern. A large truck idled on the pier and workmen were transferring pallets of food from the truck to a hook dangling from the end of *Knorr*'s crane. A few tons of frozen, canned, and dried food would be required to sustain the ship for the next 60 days. The steward scurried about. He had his hands full, taking inventory of everything coming aboard, with the additional task of finding a place to store it all.

As Amy and I lifted our personal gear out of the taxi and walked toward the gangway to report on board, voices echoed from inside a dumpster a few feet away. Once aboard *Knorr* I looked down from the higher elevation of the deck, and could see right inside the large metal container. Two men were lifting a computer monitor out of the trash.

A man in oil-spattered denim coveralls came up beside me. His pasty white skin advertised his position in the engine department, a job spent mostly in the bowels of the ship far from sunlight. "They've been doing this all day. We throw junk in the dumpster and before it hits the bottom these guys are hauling it out. I can't imagine what they're going to do with that stuff. The monitors are burned out. They don't even work."

"Durban's version of recycling," was all I could say, as I watched two small men heft an empty wire spool over the lip of the rusted container to an eager friend on the outside. When I turned back to introduce myself to the man in overalls, he was gone.

As Pieter our cab driver had said, "Poor people are desperate people." What did these men intend to do with those defunct CRTs and all the toxic materials contained in them? It was one of the many questions about life on a Durban dock that had no apparent answer.

The travel industry liberally scatters the word "exotic" in its advertising when it wants to lure visitors to faraway lands. The dictionary defines exotic as being "from another part of the world," but nowhere in the definition does it promise that it is a better part of the world. I have learned that exposure to distant places comes at the price of lost innocence. My postcard images of paradise are regularly shattered when I'm given the opportunity to view the scene from the other side of the camera. My image of Durban had been shattered. I could only hope that the rest of South Africa remained somewhat intact.

After we got all our bags aboard, Amy and I tried to find the way to our cabin. The now-familiar smell of a working ship filled my nose. I didn't have enough experience to say that *Knorr* smelled different than *Oceanus*, but I was willing to bet any one of the long-time engine crew would know. I was supposed to be leading Amy around, but the convoluted passages aboard *Knorr* had me lost in a matter of minutes, and I had to get directions from some of the crew along the way. I dragged our duffels up another deck and looked back at Amy. With backpacks over each shoulder and a third bag in one hand, she looked like a pack mule. Another thing I noticed was that she had no problem maneuvering through the halls and ladders. She was back in her element.

We would not be making the dead-head leg up to Kenya, so

after finding our bunks we didn't unpack but rather quickly threw our stuff in a closet and headed down to the lab.

This was turnover day, the day most of the ship's science party and crew changes took place. There was a lot of activity, but strangely almost none of the typical ship's banter among the people we saw. I mentioned it to Amy but she was more interested in checking the status of her newly delivered equipment and hadn't noticed anything odd.

After inspecting all of the incoming gear, and lashing it down securely against any potential bad weather, I took a break and followed the smell of freshly baked pastry to the galley. A mound of chocolate chip cookies sat undisturbed on the counter. I put one on a plate, poured a glass of milk, and moved to sit among what, at first glance, were deserted tables. It was only then that I saw a woman sitting alone in the room.

She was about 50 years old and had long blond hair that hung over a cold-looking cup of coffee. Beneath her glasses her eyes seemed to be staring at a point on the other side of the steel bulkhead. "Mind if I join you?" I asked her, and sat down to introduce myself.

Her name was Linda and she was the ship's medical officer. Normally the role of ship's doctor falls upon the captain, but given the remoteness of the previous leg, and the hostility off the coast of Somalia where we would be going, WHOI had decided to place a nurse practitioner on *Knorr*. It was a comfort knowing she is aboard. After fire, a medical emergency at sea is a sailor's biggest concern. Even a sinking can be less risky. If the ship goes down, modern communications and survival rafts make it only a matter of time before help arrives. It can take days, however, for a ship offshore to get close enough to the coast for rescue helicopters to airlift a person to a facility capable of treating a severe injury or illness.

Linda told me this was exactly what had happened on *Knorr*'s previous leg, and it was the reason for the subdued demeanor of the crew. "Randy, one of *Knorr*'s ABs, came to me with a headache.

The next day it was worse. He complained of facial numbness and was talking to me in the infirmary when he suddenly collapsed. The next thing I knew his respiration stopped and he was comatose. I called for help and started artificial respiration. Ultimately I got a bag on him. You squeeze the bag every 10 seconds to force air into the lungs. Randy was stable, but we had to breathe for him. At the time the ship was almost 1000 miles from Cape Town."

"That's days away!" I said, amazed at the story I was hearing. "You pumped the bag that long?"

Linda finally smiled. "Every member of the crew took turns sitting with Randy in half-hour shifts. They talked to him, encouraged him, and squeezed that bag every 10 seconds for three days until we were close enough for a helicopter to take him off. He was pronounced brain dead from an aneurysm the moment they got him to the hospital, but I don't think one person felt they wasted their time with him."

Leaving Linda there with her thoughts, I thought about what it must have been like racing back to Cape Town under those circumstances. These people lost a shipmate and a friend. The tragedy demonstrated to me the responsibility people feel for one another at sea that they may not feel ashore. I realized that there would be times in our future when Amy would go to sea and I would stay behind. As terrible as were the events of *Knorr*'s last leg, there was comfort in knowing that Amy would be sailing among people who would go to heroic lengths to keep her safe.

10

SEVEN-TENTHS OF THE GLOBE IS COVERED BY WATER. Scientists whose job it is to study these seven-tenths find themselves traversing the other three-tenths with some regularity. Travel, it turns out, is one of the great perks of oceanography.

Oceanographers, with a little advance planning, can take advantage of their globetrotting lifestyle to experience exotic cultures and locales by detouring their travel at the beginning or end of a cruise. There is a small bookstore where we live that has a contest every month. The winner is the person who can produce a photograph of the store's sales bag taken the furthest from the store. In the oceanographic community of Woods Hole, taking a picture of a colleague holding a bag under a palm tree on the remote island of Venua Levu is no guarantee of winning.

Many ports of call for research ships are sequestered in the seedier sections of cities. Some are industrial terminals handling freight, crude oil, and bulk materials, while others support diminishing fishing fleets. Less often, an oceanographer will meet up with her ship to find it berthed next to a luxury cruise liner in a city frequented by tourists.

Regardless of the port conditions, Amy and I were not about to fly halfway around the globe to Africa and return with only memories of a decaying port in Durban.

All arrangements for our safari had been made via email with

Jeremy, guide and owner of the tour company. He kept assuring us that we could be very flexible in our itinerary once we were there. I thought this all well and good, but wondered how the other participants of our tour would feel if we wanted to changed the plans.

It turned out I should not have worried; Amy and I *were* the only participants, though not from any lack of marketing on Jeremy's part. By design, it was a personal five-day tour of Zulu Land and the Umfolozi game reserve. Teddy Roosevelt probably didn't find this level of service unusual during his expeditions, but we were not accustomed to having a personal driver/guide/lecturer at our beck and call 24 hours a day. As I unpacked our suitcase, the bookstore's bag was uncovered from beneath a pile of socks. With the confidence of a politician in an uncontested election, I pulled out the red and white plastic bag and placed it next to my camera.

The next morning Jeremy met us in front of the hotel for our first day in search of big game. His white van was painted in zebra stripes, which looked misplaced in the urban jungle of Durban. Amy and I peppered Jeremy with questions and found that he was not only adept at identifying flora and fauna, but had a deep understanding of the political and cultural history of this country as well.

After 15 minutes, the high-rise buildings of the city shrank beneath the trees as we sped east in the van. The highway system in South Africa is as good as any found in North America. It is so good that for many South Africans it provides the most direct route between villages, and many people take advantage of it. The problem from what I could see out the van window was that the majority of South Africans were not in cars. Long lines of people filed along the asphalt. School children in matching blue uniforms carried books on their heads as trucks speeding a few feet away kicked up clouds of red dust.

"We do have a lot of pedestrian fatalities here," said Jeremy matter-of-factly.

Jeremy had a slight Afrikaans accent, the product of his Dutch roots. He was none too tall, with a thick build. His reddish beard and mustache were set off by a green bushman's hat. Khaki shorts and shirt and green knee socks sprouting from leather hiking boots completed his tour guide uniform. Jeremy would not have looked out of place in an old sepia photo, one foot up on a dead lion, shotgun resting on his hip.

As Jeremy propelled our air-conditioned, four-wheeled zebra down a modern stretch of pavement, I realized that this adventure into the bush would be a tightly choreographed one, despite his best efforts to make it feel like a trek into the heart of darkness.

As if to reinforce the aura of rustic bushwhacking, Jeremy reminded me that tonight we would be sleeping in tents.

"Your website didn't mention *tents,*" I protested.

Jeremy did not answer, but I saw his lips in the mirror stretch into a smile.

Paying to see all of the same animals that we have seen in the zoo and on countless hours of National Geographic television might at first seem redundant, but there is some primal euphoria that comes with stumbling upon these same creatures in the wild. Capturing a close-up photo of a white rhino is nothing more than the modern equivalent of stalking one and clubbing it to death for food. "See honey! Look what I brought home on film!" can be as rewarding as bringing a wildebeest back to the village for the tribe. There is also the matter of perspective. A half-hour spent watching a lone dung beetle roll a ball of dung 20 times bigger than herself down a dusty path to her nest made me want to call my mother and thank her for all the diapers she'd changed when I was a baby.

We turned onto a dirt road that would bring us to our first night's camp. An earthy smell surrounded us. Dirt lifted up by the tires swirled into the open windows. It was hot, but neither Amy nor I wanted to lose the sensations of this place by turning on the air conditioner. The van continued along the narrow path, and suddenly we came across a spider as big as a fist. It had strung its web across the width of the road. As the last remnants of spider silk blew from the windshield Jeremy pulled the van through a gate made from two saplings bent toward each other and joined at their tops.

"Here's your tent," said Jeremy, as he stopped the van and got out.

We walked through the flap. It *was* a tent, technically; canvas walls and roof. But it was like no tent I had ever used. In the center of the room stood a four-poster bed with turned spindles of dark mahogany. Matching nightstands and bureau sat on an ornate Persian carpet. I flopped down on the thick down comforter and stared up at a slow-moving ceiling fan. Amy walked around the room, pacing off its scale and getting a sense of the furniture details with her hands.

"If I had known this is what camping was like I would have joined the Boy Scouts," I said.

Despite Amy's poor vision, even she could see the opulence of our accommodations. She found her way behind the bed's headboard, which acted as a partition to another room. "Do you know there is tiled bathroom back here with a marble lavatory?" Her voice echoed in the space. Before I could react to her statement, she asked, "Are you sure you checked all my shipping containers that came aboard *Knorr*?" I could hear the sound of water running in a sink.

"Three times. The first time I counted six, the second time there were half a dozen, the third time there were three pairs," I replied.

"I can't help thinking I'm forgetting something," she said, coming back into the room.

"Well, until we and *Knorr* get to Mombassa, there is not much you can do, so relax, enjoy the trip." This was directed up to the ceiling fan spinning over the bed.

The next morning we awoke to find a small heard of gazelles grazing outside our door. The air was perfectly still and the dust from their hooves hung in the low light. We crouched low in front of the tent flap. Amy brought her monocular to her eye and I whispered directions for her to turn until the animals came into her view. We watched them quietly for half an hour. Looking at Amy, I sensed that stress of the cruise logistics was finally beginning to wane.

Over a gourmet breakfast of poached eggs, cheeses, and a variety of meats I could not identify, Jeremy told us that the only place big game animals survived in South Africa was the game parks. In fact, most Africans had never seen a live elephant. Even though the cost to enter the park was only pennies, it appeared few South Africans cared to see these wild creatures. This seemed all very tragic until I realized that, growing up in Boston, I had never been to the top of the Bunker Hill monument.

After a few days in the van Jeremy surprised us with a boat trip on the Hluhluwe (pronounced: Shlew-shlew-way) River. Since the next five weeks would find us surrounded by water, we were hoping not to see much of it during our tour of the bush, our logic being that an infusion of arid plains into our subconscious would allow us to better withstand the terrestrial deprivation we were about to be subjected to. But we went along.

The boat was a large open skiff with bench seats running along both sides. It was beached on a muddy riverbank where a narrow plank leaned against the bow, presumably to facilitate boarding. It looked like an agility obstacle at a dog show. The river flowed gently around the boat's stern, though from where we stood the water looked to be more marsh than river.

We were joined at the bank by a vanload of people from another tour company. This group of six, a family from England, seemed more excited about the beer that the boat operator was loading aboard than about the prospect of seeing wildlife.

Amy took a tentative step onto the plank, then another. It sprang up and down every time she lifted a foot. I had to guide her using only my voice because there was no way the flimsy board would hold both of us together. Like a tightrope walker she held her arms out for balance and effectively, if not gracefully, jumped into the boat. After everyone was seated, the captain of our small craft pushed the bow off the bank and hopped in. Two outboard engines sputtered to life with a coughing fit violent enough to scare anyone with emphysema. A quick shudder was transmitted up from under my seat as the reverse gear of the transmission was engaged and the boat began to move backward.

Forty feet from shore the current captured our boat and pushed it sideways, the bow still pointed at the riverbank.

"Shit," said the captain. At least I think it was "shit"; he spoke in Afrikaans.

"What's the matter?" Jeremy asked.

"No steering, can't turn the boat," the captain said, shutting down the engines. "We'll have to pole our way back to the shore."

The captain dragged two long aluminum poles from under the starboard seats and with Jeremy's help got the boat up-current to the bank where we'd first shoved off.

The morning sun was beginning to evaporate the water around us and the air thickened with humidity. With no breeze from a moving boat we were at the mercy of the heat and bugs. This was enough motivation for the English family to open the beer cooler and pass out bottles of a local South African brew, while Jeremy and the captain bent over the motors.

After 30 minutes the captain stood up with nothing to show for his labor except greasy hands. This was the signal for the rest of us to offer our own interpretation of the problem and voice possible solutions. Everyone but Amy seemed to have something to say about how to fix the steering linkage as the beer lubricated each of our tongues enough to voice an opinion.

An hour later we were gliding down the Hluhluwe; the final solution to our mechanical problem was to disconnect one engine from the linkage.

Amy reached into her backpack and removed a 10-power monocular. I wiped the grime off the lens and focused it for her. We both scanned the riverbanks in search of something of interest, finally resting on what looked like a floating log. It was a crocodile. The eyes and snout seemed disconnected, floating separately on the water's surface. Colorful birds flitted from branch to branch in the overhanging trees as we kept our distance from a group of rotund hippos wallowing in the shallows. I turned Amy's head as she held the monocular to her eye. It was like panning a camera around to capture a scene, and was an effective way to bring the images within the scope's narrow field of view.

Fortunately the sights were not the only way to appreciate this expedition; we were constantly enveloped in a natural symphony. Birds had songs that were as strange to us as the Zulu language. The tendrils of odors carried on currents of air reminded me that we were far from the fresh-cut-grass comfort of home. To have an intimate experience with the African continent we had to open up all of our senses and allow its richness and diversity to enter us. Forgetting to take a deep nose full of swamp air would mean missing an important part of what made this region unique.

Three days later Jeremy dropped us off at our hotel in Durban, and back into the frenzy of urban life. Amy's thoughts immediately turned to our reason for being 10,000 miles from home. Within minutes of entering our room, she was communicating with Woods Hole to verify that there had been no travel delays for the rest of the science party meeting us in Kenya. She also checked the arrival status of the ship, but I didn't know what she could possibly do if it came into port later than planned. When Amy was satisfied that she had the latest of information, and had repositioned all of the little pins on the imaginary map in her head, we went off for a last walk around Durban before the next day's trip to Nairobi.

On the sidewalk in front of the hotel Amy tapped her red-tipped white cane along the uneven pavement. She attracted the attention of almost everyone we passed. When we'd first arrived in Africa we could not understand why everyone was staring at us, but I finally realized that we had not seen another disabled person since entering the country.

The next morning, while we were standing in front of the check-in counter of Kenyan Airways, I noticed two things. One, there was nobody else waiting in line, and two, the ropes set up to define the nonexistent lines did not match up with the counter stations.

Amy and I stood dutifully waiting to be recognized by the next available agent on the side of the rope that best seemed to approximate the agent's station. The two neatly dressed men behind the counter shuffled papers and scurried back and forth, never once looking up at us. Either they were truly busy or they had mastered the art of looking busy.

Within a few minutes people began lining up, not only behind us, but beside us on the other side of the rope. I said nothing.

Amy had enough vision to see what was happening. "You should tell them that the line forms behind us," she whispered in my ear.

"I don't think that would be very polite. I have the feeling that's not how things are done here," I whispered back.

Call me a coward, but my immersion into the local culture did not include getting into a fistfight over queue etiquette. In this case those on the other side of the rope felt that they were in the correct line, and after watching one man push his bag with the tip of his shoe closer to the counter than mine, I realized that he assumed he had the right to be waited on next.

More and more people had begun to arrive at the counter, randomly selecting one line or the other. By now eyes on both sides of the rope were darting back and forth, estimating how this was going to play out.

As expected, things degenerated into a lot of pushing and tugging. Not having the experience or tenacity of our fellow travelers, Amy and I were swept aside like pieces of flotsam when the agents opened for check-in. When we finally washed up at the counter all of the computers were down. What the agent could tell me with some certainty was that Amy had a reserved seat and I did not. He added that if I would wait "over there," he would let me know when he had resolved my problem.

"Over there" turned out to be a seat next to the information booth, which was adjacent to a set of doors marked "All Departing Flights." Amy and I sat down. Amy sat with the confidence of someone guaranteed a seat on the next flight, but I couldn't take my eyes off the agent. He was on the phone and I wondered if he was really doing all he could to get me a seat, or was just planning a tryst with his girlfriend.

Confined to the ticket area with a potential transportation crisis looming over my head, I became distracted by person after person attempting to get to his or her flight through the departure doors, only to find them locked. When it seemed that there was no way to the gates, each person approached the information booth where a large, lethargic woman sighed in exasperation and instructed the flyer to enter through the "Do Not Enter – Exit Only" doors at the other end of the terminal. This routine persisted for well over an hour, involving well over a hundred people. I am sure that this had been going on for several days prior to my arrival and would probably continue for days. At no time did the woman in the information booth try to alleviate her obvious displeasure at having to tell EVERYONE where to go by placing a simple sign on the inoperable doors. Maybe it was just job security.

After nearly an hour I was no closer to getting a seat on the flight. "You'd better plan on leaving without me. I'll catch up, just don't let the ship sail without me," I said to Amy, who had been waiting with unfair calmness, given my situation.

"I'm sure there are more flights later today, and the ship doesn't leave for another 36 hours," she said. I was a bit put off by the fact that she didn't volunteer to stay with me, but then again she was a lead scientist on this cruise, and I was only a sacrificial research assistant. At such moments wedding bands carry little clout.

But before I could protest Amy's abandonment of me, the gate agent walked up to us, tickets fanned out in his hand. "Here you go Sir, your flight will be boarding in a few minutes." I was tempted to try to walk through the door marked "All Departing Flights" for no other reason than to annoy the woman in the booth, but not wanting to push my luck I picked up our carry-on bags and headed to the Exit door.

11

AMY HELD MY ARM as we strolled the grounds of our Mombassa hotel. Holding on to me was equal parts affection and necessity, since she found it easier to leave her cane behind and use me as a guide. Amy had little trouble navigating familiar territory like her office, or our house, but in the evening light, the twisting gravel paths presented too much of a visual challenge. In addition, the low-hanging branches of the native trees added obstacles that a cane would never detect.

We rounded a corner of the building and spotted two men walking toward us along the gravel path. They each carried a drink in one hand and looked at the world through mirrored sunglasses. One man was in his forties and had a stocky build, his broad square shoulders incongruously stuffed into a tropical print shirt. A full head of salt-and-pepper hair with matching mustache could not conceal an array of small scars scattered about his face. It was obvious that he was American by the way he swaggered—it was as if he was trying the country on for purchase. The British walked this way until just before the end of their empire.

His partner was much younger and taller, with a slim build and boyish features that he tried to compensate for by walking—well, there was no better word than "mean." He wore a sport shirt and blue jeans that were pressed so sharply the creases would probably draw the blood of whomever they brushed against. This boy looked as if he was out on a weekend pass from some marine boot camp.

As I described the two men in front of us to Amy, we both reached the same conclusion: these must be the two men from the security company contracted by WHOI. When we got within a few feet, Amy asked them directly.

"Are you two the security team that's coming aboard *Knorr*?"

The older one lifted his sunglasses and glared at us as if we'd just discovered the launch codes to a nuclear missile.

"Why do you want to know?" he replied in a tone that suggested we had no business asking. He seemed genuinely surprised that anyone could have identified them, cleverly disguised as they were. He swirled the ice in his glass, another indicator that they were probably American, since ice cubes were a rare request outside of North America. "Are you two scientists?" he continued, as if to show he was as good at the guessing game as we were.

After a few minutes, all we'd gotten were questions in response to our questions, and no confirmation of our original assumption. I was thinking of forcing the issue by asking them if they were in Mombassa for the Gay Pride festival, but decided it wasn't worth the risk of losing my significant dental investment.

Amy and I walked away, amused at the non-conversation that had just taken place. "I'll bet you a nickel we'll see both of those guys on the ship tomorrow," she said to me. Amy never bets on anything less than a sure thing, and even at that, never more than a nickel.

"I hope you're wrong," I said, looking over my shoulder at our potential protectors. I couldn't feel too confident in a guy who tried to blend into the fabric of Kenyan society by wearing a Don Ho shirt over pasty white arms.

A few hours later we were ready to report aboard *Knorr*. She was moored against a crumbling concrete quay next to a ferry that shuttled people across the river dividing the city. The ferry was a much larger version of the *African Queen* and probably of the same vintage. I almost expected to see Humphrey Bogart step out of the bridge with an oily bandana tied around his neck. Two tall black stacks with fluted tops directed grey smoke into the sky. A seemingly endless line of people waited in the hot sun to make the trip. They crammed bicycles, donkeys, chickens, and wooden carts onto the deck for the crossing, which wasn't particularly long, a half-mile at the most. But my guess was that given the conditions aboard the ferry, it was a risky undertaking. There were probably only enough life jackets aboard for the captain and one mate.

On *Knorr* we found the accommodations, unlike those on the adjacent ferry, to be quite spacious, especially since we were carrying less than half the regular complement of 25 science personnel. This meant that each person had a cabin to him- or herself. Amy and I shared one, which I believe violated the letter of WHOI's rules forbidding persons of opposite sex to cohabitate, but unlike that time aboard *Oceanus*, Amy hadn't actively sought other sleeping arrangements. On British ships the non-cohabitation rule is strictly adhered to on the grounds that even married couples should not bunk together for fear that it would depress the rest of the crew who "weren't getting any," or whatever the British equivalent phrase would be.

At 279 feet, *Knorr* was the largest type of research vessel in the University National Oceanographic Laboratory System (UNOLS) fleet of 29 ships. Considered "Global" class, *Knorr* was capable of extended service in any ocean that didn't freeze over with ice. She shared the color scheme of the other two large vessels operated by WHOI, a dark blue hull and pale green superstructure. Inside, the volume that this hull encapsulated seemed immense. Even with our previous orientation aboard her in Woods Hole, and the short time aboard in Durban, I still felt myself lost amidst the many decks and spaces. To make things more confusing, *Knorr* had 'tween decks. These are decks that are not lined up with adjacent decks, but fall somewhere in between. Amy handled this maze with the guile of a lab rat, and before long she was giving me directions back to our cabin.

The next day at midmorning, preparations began for our departure; the gangway was brought aboard, and last-minute deals were struck between members of the crew and souvenir vendors on the dock. Money was handed down over the rail as shiny wooden animal carvings were passed back up. *Knorr* moved away from the dock accompanied by a long blast of her whistle. The sound reverberated in my chest. I found a patch of shade near the starboard life raft and watched as we motored slowly toward the mouth of the river. Ferries lumbered across the narrow harbor entrance as we picked our way between them. They sat so low in the water that it would probably have taken only an extra chicken or two to make them sink. Two of these identical boats crossed us in mid-channel, headed for opposite banks. For a split second their bows seemed to touch as if each was bumping against a mirror. But like Alice passing through the looking glass, they continued on, the silhouettes merging to form one ship before dividing again like a reproducing amoeba.

I was getting into the rhythm of the port when Francis took up a position next to me.

Francis was one of the rated seaman aboard. The letters of his rating, AB, stood for Able Bodied, a position that goes back hundreds of years in the maritime tradition. ABs have completed more training than nonrated crew and can serve on the bridge alongside the officer on duty.

Francis was in his fifties, a big guy, a few pounds shy of being fat, with a cherubic face that was always smiling. He was also the ship's newspaper. He knew everything that was going on, both aboard and ashore. "So did everyone escape Mombassa without incident?" I asked, to make conversation.

"Mombassa yes, I think everyone learned their lesson in Durban."

"What happened in Durban?"

By the arch in his eyebrows it was obvious he had a juicy story to tell.

"Two scientists had an interesting evening the night before we left," he said, his voice hardly containing his excitement.

Over the next five minutes, Francis related an incident involving a man and woman who had made the transit up from Durbin. Both of these people I subsequently found to be intelligent and well traveled, which in light of the remainder of this episode attests to the judgment-numbing effect of alcohol. It seemed that while walking back to the ship after some late-night imbibing at a local pub, they heard the call of a deserted Durban beach beckoning past the boardwalk. After stripping down and frolicking in the surf, they returned and found, to nobody's surprise but their own, that their clothes, money, and passports had been stolen. The man found the situation quite distressing, while all the woman could do was laugh, and since the only thing left in their possession was a sense of humor, this seemed to be the thing to do. They explained their predicament to a wayward drunk, who, after a period of uninterrupted listening, replied, "Yup, you-a screwed mon," before staggering away shaking his head.

The pair walked into the lobby of a nearby beach hotel dressed just as they had been when they arrived into this world. They were saved from additional embarrassment by an unruffled bellman who came to their aid with towels, which probably made them not one bit less conspicuous amidst the marble and potted palms.

Remembering that *Knorr*'s offgoing captain was staying at that hotel, they proceeded to knock on his door, forgetting that it was three in the morning. The captain felt only marginally responsible, since he had already signed off the ship, and without ever taking the security chain off his door, tossed them a minimal amount of clothing and enough money for a cab back to the dock.

As Francis finished his story and we both watched Mombassa recede in the haze, I looked down at the deck below, saw the male scientist of Francis' story, and tried my best not to picture him naked walking through a hotel lobby.

It would be five days of steaming north along the African coast to the work area in the Gulf of Aden. This transit period provided us with ample time to set up and test instruments, secure computers, and establish a plan for the work site. The latter task involved

Amy and the other PIs, principle investigators, standing over a chart of the region and positioning the sampling stations for the most economical use of ship time. The result was an uninterrupted jagged line of dots traversing the Gulf of Aden, dots that became more tightly spaced the closer they got to the Red Sea.

This exercise in strategic planning seemed to rival the invasion of Normandy in complexity. Each scientist had an opinion regarding the nuances of this yet-to-be-mapped outflow of Red Sea Water, and much discussion was dedicated to the fine details of where to place stations and on whether to start a particular line from the north or the south.

To my untrained ears all of this debate seemed trivial, until I remembered that the cost to operate *Knorr* was about $25,000 per day. This was not to say that Amy could peel off hundred-dollar bills on the bridge and expect extra days if she needed them. The entire cruise was tightly scheduled. In the end, even if Amy's grant could afford a few extra days, she would not get them because another scientist was eagerly awaiting our return so that he or she could have the opportunity to spend grant money on this floating laboratory. Conversely, if Amy should pull into port a few days early, there would be no rebate check forthcoming in the mail. The whole system operated under a strange set of accounting practices, but it worked well at providing equal shipboard access to researchers around the United States.

Making the most efficient use of ship time, and trying to predict such unpredictable contingencies as days lost to bad weather and equipment failure, is key to the success of any seagoing program. The fact that none of these skills is taught in any oceanography course makes the process all the more interesting. For the graduate students along with us on this cruise, involvement in the planning of the station work was a valuable opportunity to learn practical lessons they would never get in the classroom.

The next morning I opened my eyes to see the low light of our second sunrise at sea filter through the porthole curtain. The cabin was cast in a mustard yellow glow from the light reflecting off the painted surfaces of our steel bunk beds, wardrobe, and writing desk. A small stainless lavatory hung from one wall and there was some comfort in knowing that any dried toothpaste in this sink was either Amy's or mine. A door at the far corner of the room lead to a head shared with the adjacent cabin, a compromise that required us to keep it in some minimal level of cleanliness.

As usual, Amy was showered and ready for work before I'd even had a chance to wash the sleep from my eyes. As she walked out of our cabin she gently reminded me that we had a security meeting in the ship's library after breakfast. The snooze alarm on my clock went off for the third time. I dragged myself out of the bunk, splashed cold water on my face, and pulled on yesterday's socks. Before walking out the door I noticed than Amy had left her walking cane on the desk, confirming for me that after only a few days she had a good feel for the layout of the ship.

The mess deck looked like an ordinary, small cafeteria except that all of the furniture was bolted to the floor. Scrambled eggs were heaped on a plate and handed to me from the galley pass-through where the steward was assembling breakfast orders. I found an empty seat at a table where a few of the crew were already eating. From a small wooden box secured to one end of the table someone pulled out a bottle of catsup. This little box was an informal log of the ship's past ports of call. Looking through it I found hot sauces from around the world: Satan's Fire, Aunt Mammie's Pepper Potion, and a homemade Caribbean-island concoction packaged in an old Coke bottle. *Oceanus* had a similar assortment of spices, and one could easily get the idea that ship's crews are fueled on Tabasco.

Ben, one of the oilers from the engine department, reached past me and plucked a potent-looking sauce from the box. Splashing a torrent of red liquid liberally about his eggs, bacon, sausage, and pancakes, he continued his conversation with Jerry "Boats," the ship's boatswain.

"I don't think about it much," Ben said between mouthfuls of reddened pancakes. "There are lots of ships out here and it's a big ocean. The odds are with us."

"That's the problem," replied Boats. "Where we're going the ocean gets pretty damn small and we'll be out of traffic most of the time. Close to shore once in a while I'll bet."

"If that's the case I hope those two security dudes brought some big guns," Ben replied.

Four months ago the concept of piracy had seemed remote, just another contingency to plan for, like spare parts for equipment or vaccinations for tropical diseases. The talk of guns and a recent incident in Durban had me thinking more seriously on the matter. I wasn't nervous, but it was all easier to dismiss in the comfort of Woods Hole than here, with the coast of Somalia a scant 100 miles to port.

Two weeks earlier, after departing Durban for Mombassa, *Knorr* had been required to return to Durban when a small group of South Africans was found on board shortly after the ship set sail. By chance, the three men were discovered hiding in a deck locker when *Knorr* was only 30 minutes out of port. A Kenyan Port Authority vessel was called to retrieve the men and they surrendered nonviolently to the ship's crew.

Things could have been much worse. If the men had spent any more time in the locker they could have suffered serious consequences from the heat. *Knorr* could also have been subject to significant delays had the stowaways been discovered much later in the voyage. There were other concerns; men who stow away on ships are usually desperate characters. In most instances the conditions that stowaways are leaving behind are so bad that they do not even care where the ship they are hiding on is headed. Under such circumstances a confrontation aboard *Knorr* could have easily become violent.

But for all the talk of piracy being bandied about the table, no one, not even the seasoned seagoing people here, could recall any serious threats to ships they'd been on. Despite the recent attack on the U.S.S *Cole* in Yemen, and this talk of guns, we seemed so removed from any danger that it was hard to take the possibility of an attack seriously. After all, who would want to steal anything from a research vessel? There were container ships crisscrossing the Gulf with millions of dollars worth of TVs and DVD players ripe for the picking. The general consensus around the table was that *Knorr* was not worth the bother when a more lucrative target might be just over the horizon.

After breakfast the science party began to assemble in the library, a small L-shaped room lined with shelves containing scientific reference books and an assortment of paperbacks donated by previous travelers. The electronic age had necessitated the addition of a computer to the ship's library and it was rare to find the chair in front of it empty. The crew member sitting there now noticed the ever-increasing number of bodies filling the room, and not wanting to sit in on any meeting he was not required at, logged off the computer and left.

After everyone was accounted for, the captain entered, followed by the two men Amy and I had met in Mombassa. I elbowed Amy in the arm and whispered in her ear, "It IS them. I owe you a nickel." The pair had ditched the tropical print shorts in favor of khakis. The captain introduced them and explained that they would be on alternating 12-hour watches as a dedicated set of eyes in search of potentially dangerous situations. They would also be scheduling training sessions for scientists and crew in methods of self-defense and securing the ship in the event of an attempted boarding.

Jim was the older one and Vince, when he finally did speak, seemed a lot less mean than my first impression of him had suggested. They talked to us about safety precautions and the role the science party would play in the event of a pirate encounter. It turned out our part involved a lot of running and hiding while we left the attackers to Jim, Vince, and the ship's crew. The crew of *Knorr* had been undergoing separate training during the transit on the art of repelling boarders and the rooting out of stowaways.

To help everyone understand the severity of different situations, Jim had devised a rating system for escalating levels of threat conditions: Alpha: a suspicious boat is seen in the area; Bravo: pirates are exhibiting hostile behavior; Charlie: pirates have boarded the ship. Jim and Vince constantly talked in code: "In the event of condition Charlie all crew will retreat below and lock down until the SITREP reverts to Bravo."

I got the feeling that if someone ran around the ship yelling "PIRATES! PIRATES!" Jim would have no idea what the problem was.

With nobody having a desire for posthumous medals of heroism, there was no objection to Jim's plan to muster the science party in the main lab in the event of a threat. As the full details of the action plan were revealed to us, however, a few in the group took offense.

"In the event that pirates get aboard *Knorr*," said Jim in a voice that resonated about the steel walls, "all members of the science party will remain in the main lab space with the exception of the following." He then listed every female member of the science party and the two female members of the crew. When he reached the end of the list he added, ". . . and David Fisichella."

WHAT! I thought.

Jim continued. "The above-mentioned women, and David, will proceed to the lower engine room where they will be sequestered for the duration of the conflict."

The room now buzzed with murmuring and some low laughter. All eyes in the library turned to me. My mind was racing. I didn't know whether to feel privileged to be locked up in a cramped space with a shipload of women, or to feel castrated, segregated from my fellow men just as the call of battle sounded. My sole purpose for inclusion, I found out a few minutes later, was to be Amy's eyes during this ordeal. It seemed my role as guide dog now extended into the far reaches of the bilge. At that moment testosterone trumped practicality and I wished the responsibility of being Amy's eyes was someone else's. I wanted to defend the castle, not cower in the dungeon. But the more obvious question was already being asked.

"Why hide the women?" The tone left the unsaid part pretty clear: "This is the nineties, buster. You'd better have a good answer or I'm going to shove my N.O.W. card down your throat."

"We feel," Jim responded, "that pirates in this part of the world would likely treat women differently from men, and possibly use them as bargaining tools if they capture the ship."

At this I felt the need to stand up for my fellow women. "So what happens when the pirates start rummaging through the cabins and find drawers full of bras and panties? Are you volunteering to be the designated transvestite?"

This got a chuckle out of the room but not even the hint of a smile from Jim. I imagined he felt he had every contingency covered, but this cross-dressing, Merchant Marine scenario had exposed a flaw in his otherwise brilliant plan.

Things only got worse when we practiced our evacuation plan. The space we were led to was noisy and cramped. It was not an area that often saw the presence of 10 people at one time.

"I can't stay in here for very long," came a voice from the crowd.

"You won't have to," said the chief engineer, who was escorting us. He lifted the rubber mat covering a deck plate. "You're going down there."

As the chief raised a metal door in the deck, we all peered down into the dark void. "There is no way I am going down there," said one of the research assistants, expressing a feeling probably shared by more than a few others. Reluctantly though, she took her place in line. She probably realized, as I had, that this was an exercise we had to pass and it was best to get it over with.

The chief handed us a few flashlights and we took turns stepping into the hole. The space was too small to stand up in and only about 10 feet by 10 feet. Surrounded by bare metal, our voices echoed in the confined area

We crawled out of what many described as our communal coffin, and walked silently back up to the lab.

Once in the lab, we were divided into groups of six. The captain had decided that all members of the science party would be given some basic lessons in self-defense. Jim brought us into a small, unused room on a lower deck. In the next hour he went over a half-dozen moves we could use on attackers to break away from being held, to stun them, or to disarm them. Very lofty goals, I thought. Like the safety announcement given on aircraft, these lessons seemed only a diversion to make me feel empowered, without really being intended to save my life. As I stood there, half listening to Jim, I thought back to those preflight briefings and couldn't remember when knowing how to use the life rafts had saved anyone in a plane crash.

"OK, David, now I want you to do the move I just showed you on me."

Jim's voice pulled me out of my life raft thoughts.

"Don't hold back," he continued. "Do it as hard as you can. I need to know that you can do it right,"

Jim was sitting in a chair. What I was being asked to do was come up behind him and strike down with both hands in a chopping motion to the sides of his neck. In theory, a hard blow to this area collapses the carotid artery that carries oxygenated blood to the brain. This sudden loss of blood flow temporarily renders a person unconscious. My first problem was finding his neck. As best I could tell, he had none. The second, greater obstacle was to muster enough force to make the move work. As hard as I tried, I found it very difficult to hit another person; that this person wanted to be hit did not make it any easier.

I raised my arms above my head. My hands dropped like lightning bolts, but in the millisecond before they landed on him, the muscles of my back tightened up in an involuntary contraction, slowing things down enough for the hit to be ineffectual. My body, it seemed, was programmed to recognize the absurd and act against it.

I was not the only one suffering from this antiviolence virus. Jim had to ask each of us to repeat the procedure until we were so tired of him saying “hit me harder” that we wanted to kill him just so he would shut up. In the end I lost track of the blows that Jim took to the neck, arms, head, legs, and torso. We made him black and blue and he seemed the happier for it.

12

KNORR WAS SAILING NORTHEAST TOWARD THE EQUATOR, into a southwest flow of water known as the Somali Current. Part of a larger surface circulation of water that rotated around the Indian Ocean, the Somali Current contributed to the huge Indian Ocean Gyre in much the same way that the Gulf Stream impacts the North Atlantic. The main difference was that in the southern hemisphere the general motion of large-scale circulation was reversed.

If the Somali current was impeding our progress, it was not obvious to me, as we seemed to be effortlessly gliding through mile after mile of calm water. The current was there, however, slowing us down by two to three knots, eating away at valuable ship time, an unseen force trying to keep us from our destination. *Knorr*'s engines, with a combined 3000 horsepower, made the current a navigable obstacle. In the days of sail, boats could not dash headlong into this rush of water but had to find ways to use the surface currents to their advantage. By riding the ever-present circular flows as on a highway around the world's oceans, ships of 200 years ago would go with the current, not against it, even if that meant sailing much longer distances.

The importance of understanding ocean currents remains as true today as it was when Benjamin Franklin first mapped the Gulf Stream. The difference now is that, in addition to providing an advantage in trans-oceanic commerce, our present insight into ocean circulation is leading to a better understanding of our planet's weather, climate, and biodiversity.

In the past, oceanographers charted the movements of currents by piecing together small sections of surface circulation taken by sporadic *in situ,* or "natural place," measurements. The result was a mosaic of patchy observations that may or may not have been an accurate reflection of what was happening in the larger area of the ocean.

Today, satellites can instantly map surface currents with great accuracy. But even more than the accuracy they deliver, these remote sensors can capture a picture of an entire ocean's surface currents instantaneously. The laborious cut-and-paste images of past surface circulation models have been replaced by multicolored enhanced photos that depict every nuance and wispy thread of a current's characteristics.

Despite these great advancements, however, measurement of the subsurface currents remains an old "poke and probe" adventure, carried out on ships in the far reaches of earth's oceans. There is no remote sensing instrument that can penetrate more than a meter beneath the sea surface. Which is why, I reminded myself, I was crouched on the upper deck of *Knorr* breathing in the humid Indian Ocean air.

My knees were beginning to ache from kneeling on steel deck plating. Ann and I were attaching water collection bottles to the rosette in preparation for the start of scientific operations the next day. Ann was a research technician from the University of Miami. Her blond hair was cut close to her head and well-muscled arms and legs sprouted from a matching pair of tattered Florida "Gators" tee shirt and shorts.

"Pull that strap out from under the frame. I'll lift the frame from here," she said to me as she got to her feet. She pulled up on the aluminum tubing and lifted the 150-pound instrument on its edge. I gave a quick tug on the strap, not wanting to get my fingers caught in case she dropped it back down, but Ann gently lowered the frame. I stood up with all 10 digits intact.

"Look at that. I've never seen a sunset quite like this before," I said, pointing toward an orange glow near the horizon on the *Knorr*'s port side. Ann looked up from a bolt she was tightening.

"Sunset was a while ago, and twilight this close to the equator is short. It almost looks like a fire," she said.

To avoid the heat of the day we'd decided to assemble the instrument late in the afternoon, and it was dark now, dark enough that stars were visible against a near-black sky. "It could be the lights of a city," I said. "I haven't looked at the chart in a while. But Mogadishu should still be far to the north."

"Or it could be burning buildings in one of the coastal villages." I knew that Ann had every reason to offer up the fire idea as a valid hypothesis; I'd been following the desperate situation in Somalia since the inception of the cruise. There had been recent reports of fighting among warlords in this part of the country, a contributing factor in the heightened piracy threat. "You know, I swear I could hear explosions all afternoon," Ann continued.

Earlier, other members of the crew had talked about seeing arcs of light like rocket trails over the coast. Not something to worry about at the time, but that combined with what we were seeing now and the fact that Vince was perched on the deck above us, binoculars pointed to the west, gave me an uneasy feeling. The images of the body of a U.S. Blackhawk helicopter crewman being dragged through the streets of Mogadishu, and the violent acts described in the printed pages of magazines, suddenly lost their insulation of time and space. The war in Somalia was now within spitting distance.

After leaving Kenya, *Knorr* seemed a safe haven, floating on a liquid barrier separating us from such lawlessness. There had been few other ships sighted lately, and to some our solitude brought a sense of security. Looking at the hazy orange glow in the western sky, I began to feel that was a false sense. Nothing so far had suggested that the cruise should be terminated, but then again, would there be any warning signs?

We passed Mogadishu without incident, and after five days'

sailing north along the continent's east coast we prepared to round the Horn of Africa and enter the Gulf of Aden. The Horn of Africa is not to be confused with Cape Horn, which is on an entirely different continent. This horn is bathed in tropical warmth while Cape Horn, at the southern tip of Chile, is battered by snowy gales. Nonetheless, rounding either horn can be dangerous. For this reason the captain had called a meeting with the principle investigators (PIs) of the science party.

The captain entered the lab with his chief mate close behind. They took up positions on one side of the large rectangular chart table in the room's center. Gathered around the other side were Amy and her three co-PIs. I sat at a computer station about five feet away plotting station coordinates into special navigation software that gave Amy an accessible alternative to printed charts. I was half working and half listening to what was being said at the table.

Mike, the captain, spread his arms out wide and placed calloused palms down on the table. He was of medium height with a thick, but not at all fat, build. Close-cropped red hair and an equally red, but bushy, mustache suggested an Irish ancestry, but his speech was all New England. "Amy mentioned to me today that you would like to do your test cast of the rosette in the Socotra Channel. My plan was to take the ship on the outside passage around the island and into the Gulf."

"We're thinking that this is a great opportunity to get a look at the water flowing out of the Gulf and around the Horn," said Steve, one of the PIs, unrolling a detailed chart of the area, his bear-paw hands placing a weight at each corner of the paper to keep it from curling up.

"I know this wasn't in the cruise plan," Amy added, not bothering to bend with everyone else over the chart she couldn't see, "but it shouldn't take much time to do three stations, and it will save us time from having to go east of Socotra."

More justification and thinly veiled begging crossed the chart table before the captain turned to his chief mate, a man with a blond ponytail who by appearance was young for the role, but had already proven himself capable of commanding the big ship.

"What do you think?"

"It will be dark when we're passing through the channel," he replied, tapping the eraser end of a pencil on location Amy wanted to transit. "As long as we don't draw attention to ourselves we should be OK."

The captain pulled at the corner of his mustache for a moment. "I don't have to tell you what the State Department has to say about this area, but if it is this important, I'll do it on the condition that we be ready to pull up quick if things don't look right. I'll go tell Jim. He isn't going to be thrilled about this." He smiled as he got to the last statement, as if he was going to enjoy telling Jim about our new route.

There was a visible uplifting of spirits on the science side of the table as talk turned to the details of station planning. I got up from my chair and wandered over to the chart. In front of me was an outline of the northeast coast of Africa. A pencil line ran north just off the coast and turned slightly west from our present position; a faint image of a previously drawn line continued to the northeast. The new anthracite mark bisected a series of islands that began 60 miles off the tip of Somalia and were identified collectively as Socotra. A parenthetical under the name showed that the islands belonged to Yemen. Research done prior to the cruise indicated that the passage between Socatra and Somalia should be avoided and given a wide berth by ships entering the Gulf of Aden. The line my finger traced on the chart not only ignored this advice, but seemed to tempt fate by floating us directly between the two inhospitable territories.

Later, lying in my bunk in our cabin, I asked Amy about the detour. "You seem pretty comfortable with the new track."

"Mike wouldn't go through there if he wasn't satisfied it was safe," she said.

I wasn't convinced she believed it. "How important is this side trip?"

"It's important to me," she said. "There has been very little data gathered in this area. I don't know when, or if, we'll ever get back here. I don't want to waste the opportunity." I let her answer hang there while I thought about risk versus reward.

Amy must have been thinking the same thing. "Sometimes I find myself pushing thoughts of piracy out of my head," she added. "I'm so focused on having this project succeed I'm willing to let the port office, the captain, and anyone else who wants to, do all the worrying about security. When I do think about it, I ask myself the same question, but never come up with an answer."

Adding three stations to the almost 200 station dots already drawn on the chart didn't seem to me that big a return for this investment of risk, but I was not a scientist and didn't know the significance of sampling the water here, so I put my faith in Amy and the captain and let my mind loose in the pages of a book.

It was a Patrick O'Brian novel set aboard an 18th-century British man-of-war. The captain in the story was plying the same Indian Ocean waters that we were, though the depravations of his ship's crew, including scurvy, insect-infested food, intolerably cramped quarters, and the ever-present danger of enemy cannon balls, reminded me how relatively luxurious our situation was on *Knorr*. The danger of an unseen enemy remained, however, even after 200 years. Lost in the pages of maritime history, I drifted off to sleep.

The waning light of the next afternoon found us stopped in the Socotra Channel for the inaugural scientific station of the cruise. Since it was the first time the instrument package was being lowered into the water, even the off-watch personnel were standing by to assist.

Oceanographers call this operation a cast, because this heavy instrument is lowered on a cable with a crane in much the same way you would cast a fishing line over the side. Instead of casting for fish however, we were fishing for data. This is when we'd find out if all of the delicate measuring devices attached to the carousel frame over the last five days were really going to work under water.

The first cast was like the dress rehearsal of a ballet. Everyone had an assigned role. I was on deck preparing the instruments for deployment and ready to assist in putting the package over the side; those inside monitored computers that activated the instruments and collected the data coming back up the wire. After the package was back on the deck, someone would collect the water samples and do the chemical analyses. Others stood by on deck to help control the 1500-pound "wrecking ball" as it came out of the water at the end of the cast, and was swung back aboard. And just like in a ballet rehearsal, sometimes the ballerina gets dropped.

"I can't get the LADCP to turn on," said Ann, sitting in the glow of a computer monitor and speaking to no one in particular. "I keep sending the fucking command down the wire and nothing happens." This drew the interest of half a dozen people who all had to type the command for themselves before they believed it didn't work. After that, everyone felt free to offer an opinion and try to diagnose the problem happening 1500 feet below us.

The LADCP is a Lower Acoustic Doppler Current Profiler, an ingenious device that bounces high-frequency sound waves off minute particles—i.e., tiny animals and microbes—floating in the water. The device measures the shift in sound frequency reflected back as the particles move. Just as a police RADAR measures our car's speed as we go by, this instrument relates the frequency shift of sound to the speed of the particle. Since the particle is drifting with the current, the speed of the particle is the speed of the current.

To maintain watertight integrity, many instruments use a small switch that is completely enclosed inside the pressure housing. This eliminates any possibility of a leak. This type of internal switch is normally in the open position, with the contacts held apart by the force of a magnet taped to the outside of the instrument. When the magnet is removed, one arm of the switch falls, touching the other arm and closing the circuit, thus turning on the instrument. For this cast somebody had left the magnet in place. The switch never closed and the ADCP wasn't turned on before going over the side. The rosette had to be hauled back up, 1500 feet, brought back onto the deck, and redeployed after the offending magnet was removed. Amy was taking it all in stride, as if expecting some things to go wrong and happy to discover them and get them out of the way on the first cast. For my part, I was just glad that I wasn't the one who forgot about the magnet. However, what with my constant glancing into the darkness in anticipation of marauding pirates, I was not at all confident I'd be immune to that kind of mistake.

The first cast was completed and, on the whole, the equipment and instruments all worked as expected. Ann told me this meant that some instruments were working well, some worked a little, and some not at all.

The echo sounder that we had been using with limited success to determine the water depth finally stopped functioning entirely when we sailed over 5000 meters of water. This was a very expensive, custom-made, precision devise that was supposed to not only display depth, but show us sub-bottom composition. More people gathered around it for another round of "guess the fault," and to cut the tension now building in the room, I offered a solution. "How about we put it into manual mode? Tie a line to it, drop it over the side, when the line goes slack—that's the bottom." The others in the group were not amused by my diagnostic methods and I retreated to let the experts randomly push buttons and toggle switches.

Over the next few hours we completed the passage through the Socotra Channel. *Knorr* had officially entered the Gulf of Aden and was steaming at 12 knots westward. The atmosphere was more relaxed now that we had made a few successful casts and passed around the tip of Somalia without incident.

It was just after noon and Amy and I were officially off watch. "Do you want to catch a movie?" I asked her.

"I'll be there in a minute," she responded. "I just want to look at some of these numbers." She stared at the figures on the video screen. The image was magnified about 20 times its normal size on a screen in front of her, but even at that size she had to lean close to the screen to read it. The monitor reflected her concentrating face that revealed no concern about our proximity to Somalia.

"How does it look?" I asked.

"This is raw data so it's hard to draw any conclusions at this point. All I'm doing is checking to make sure nothing looks terribly wrong."

"Are you happy with the way things are starting out?"

Amy looked around the room before answering. One big problem of vision loss is that it is easy to say things out loud without knowing who may be listening. "I'm more relieved than anything else. The anticipation of getting this cruise off the dock was starting to get to me. I'm just happy to be out here."

Amy turned back to her desk, moving her hands quickly between video magnifier and computer. By her level of energy it was obvious that she wouldn't be finished for at least a few hours, so I made my way to the lounge alone after a detour to pick up some popcorn in the galley.

With our watch ending at noon it was hard to think about going to bed. I should have been tired but sleep had not been coming easily. After hanging around the lab for a while I crawled into my bunk to read myself to sleep. By the time I actually dozed off it was 8 p.m., three and a half hours before I needed to be up again.

As we got closer to the western end of the narrowing Gulf, dark silhouettes of mountains crept in on us like an approaching squall. With land so close the details of our surroundings could be viewed through binoculars. The Gulf of Aden is bounded by deserts and covers an area about the size of California. High, brown mountains shrink to dusty sand dunes as they approach the water's edge. The contrast between the water and land could not have been greater. From what little there was to observe from the ship, it appeared that more life resided under us than existed ashore; squid, jellyfish, and dolphins could be seen swimming in the warm water. Looking out from *Knorr*'s bridge, I found it difficult to comprehend why this region should be the focus of so much conflict. I tried to imagine a life so difficult that people would say, "This is my patch of unfertile dust and I'm willing to die to keep it."

Yet there was something very special about the place. The Gulf of Aden is not only a funnel for much of the world's ship traffic; it is also the region from which man ventured forth to populate the rest of the globe, a place of importance in ancient history. Our early ancestors crossed the land bridge that existed where the Red Sea now flows, and spread across Arabia and Western Europe. What motivated them to move on? Food? Fresh water? Climate? Or maybe it's just that the sand is always browner on the other side of the oasis.

While all of this movement of mankind was taking place on the earth's surface, a movement of the earth's crust was under way deep beneath. Knorr rode almost a mile above a great rift separating East Africa from the Arabian Peninsula. This geological phenomenon is known as a triple junction. It is where three continental plates join; in this case, they were being pushed apart at a rate of about one millimeter per year. Our measurements were not directly related to this phenomenon, but in the corner of the lab a lone computer continuously pinged the bottom with high-resolution SONAR. This instrument projected a wide swath of sound to the sea floor and when the resulting echo was processed it yielded a strip of water depth over an area many kilometers wide. The resulting data display was painted with a mosaic of colorful patterns corresponding to the depth and contour of the sea floor below us, revealing a detailed map of underwater mountains and valleys. All of the information was being stored for future review and analysis by other oceanographers interested in this geology.

At the western terminus of the Gulf of Aden is Bab Al Mandeb, the narrow channel between Yemen and Djibuti where the Red Sea meets the Gulf of Aden. The literal translation of this Arabic name means "Gate of Lamentations" or, more poetically, Strait of Tears. Legend has it that women would line the high cliffs of the straight to watch their husbands sail off to war, shedding tears of sorrow at the thought that the men would never return. Like the women's salty tears, the water flowing out of the Red Sea is more saline than the water in the Gulf of Aden. This is due to the higher evaporation rate in the area surrounding the Egyptian and Saudi Arabian deserts. Because there is a higher concentration of salt in the water, it has a greater density than the Gulf Water it is flowing into, and thus, as Red Sea Water passes through Bab al Mandeb, it sinks until it finds Gulf Water of the same density. Once the Red Sea Water finds its place or depth of equilibrium in the water column, it begins to spread out and travel farther east.

Part of our mission was to track the spreading pathways and measure the speed of this water mass's flow, and finally describe how it interacted with water from the Indian Ocean.

I was off watch and climbed up to the bridge to take in the view from a higher vantage point. The captain allowed free access to the bridge as long as visitors were sensitive to the work being done there and didn't get in the way. As I reached the bridge deck I noticed there was a higher level of formality than normal. The captain and the third mate were pacing the floor with binoculars in hand while the AB on watch stood ready at the maneuvering controls.

"Come left 10 degrees," the mate ordered the AB, in a firm but polite tone.

"Left 10 degrees, aye," parroted the AB so there was no confusion with the order.

I refrained from the usual greeting and quietly took up a position out of the way. As I looked out the window I saw the reason for the formality. For the first time since *Knorr* had left Kenya, she was surrounded by other ships. Not just the lone one or two near the horizon that we had become accustomed to, but dozens of tankers, bulk carriers, and container ships racing purposefully through the water, like monstrous ants heading toward a sugar bowl. The number of ships I counted was limited only by how far I could see into the haze.

The nautical chart showed the reason for this procession. Bab al Mandeb was the place where all ship traffic coming into or out of the Red Sea converged while ferrying cargos between Asia and ports in Europe and the eastern United States. For *Knorr*'s captain, who bore the ultimate responsibility for avoiding collisions, this was far more nerve wracking than a mid-ocean hurricane.

"Slow to eight knots. No, make that six, I want to give that tanker plenty of notice of our intention." The captain's voice was muffled inside the glare hood of the RADAR display.

"Six knots," repeated the crewman as he pulled back a lever on the control console. Navigating big ships in confined spaces around other ships is an exercise in mind reading. Since ships of these tonnages can take a mile or two to come to a stop, pilots must always be predicting the future, and projecting where each of the other vessels would be five, 10, 15, and 20 minutes from the present. The science mission that took us into these confined waters made this navigation even more complicated.

The plan of Amy and her colleagues was to gather measurements across the strait, which necessitated moving *Knorr* in a cross section through the line of traffic. It had taken all the negotiating skills that they possessed to convince the captain that we must cross back and forth at right angles to the traffic, "and, oh, by the way we needed to stop in the middle to take water samples."

I continued to watch as *Knorr* threaded her way between a large oil tanker and a container ship. Glancing over the captain's shoulder at the green glow of the RADAR screen, I saw at least a dozen targets, each one representing a ship and identified with a numbered square. Each square had a line projecting out of it showing the target's direction and speed of travel. There were many more targets on the screen than I could see out the window with my eyes. It was like watching a video game in which the object was to sink oil tankers, with points accumulating in millions of barrels of crude oil. The captain played this game with a cool hand.

As the sun rose on our eighth day at sea *Knorr* began a zigzag course that would take us away from Bab al Mandeb and back eastward for the remainder of the cruise. After we gathered godzillabytes of data across the narrows at Bab Al Mandeb, the station lines began to spread out as we zigzagged north and south between the coasts of Yemen and Somalia.

We had done so many casts that the operation had become as reflexive as walking. I could feel the heat of the sun building up under my hard hat as cable payed out into the abyss. The instrument package was lowered on a thick, steel cable containing a center of conducting wire that transmitted data signals back up to the ship. I was watching the cable, which had started out vertical, move slowly closer to the ship as the angle slanted in. When it appeared as if it would contact the hull I called to the bridge on a nearby intercom.

"I've got an inboard wire angle on the rosette, it's about a foot from the hull."

"Rodger that," came the voice of the mate on watch through the speaker.

By the time I got back to the rail the wire was scraping the hull as it went down. One deck above me the winch operator was evaluating it himself.

"I'm stopping it here until the bridge can move the ship over," Francis called down to me.

A strong current was either moving the ship along the surface of the water or a sub-surface current was moving the instrument under the ship; either way, the ship and the heavy instrument package were moving at different speeds, and the ship was essentially running over the gear like someone tripping over his dog's leash. Regardless of the cause, if the cable continued to scrape the hull on its way down, the abrasive wire would cut into the steel hull like a band saw blade. Since the mate's job was only secure if he managed to keep water out of the boat, he elected to move *Knorr* so that the cable stayed off the hull. For us this movement meant that we would not be sampling the water from a vertical column directly under the ship, but rather along a wavering line that originated from the position where *Knorr* first stopped. For a very deep cast this could be problematic, but for a shallow cast such as this one, under 1000 meters, it was acceptable, especially considering the consequences.

The engines and bow thruster rumbled softly somewhere under

my feet, and within minutes the wire was a few feet from the hull. Looking over the side, I could see a mark in the paint where the cable had been, a minor scar in the service of science. Hopefully this was the worst damage we'd incur over the next five weeks.

13

U.S. State Department Travel Advisory: *U.S. citizens are urged to use caution when sailing near the coast of Somalia. Merchant vessels, fishing boats and recreational craft are at risk of seizure and of their crews being held for ransom or subject to physical harm.*

THE NOTICE POSTED ON THE BULLETIN BOARD IN THE MESS DECK was part of a security briefing circulated months ago. Tacked up next to ship schedules, OSHA safety rules, and other mundane bits of ship's business, the warning was lost among the other papers. In the three weeks since we'd left Mombassa, life on *Knorr* had settled into a placid routine. Even the weather showed no sign of hostility: sunny days, light breezes, and a tranquil sea.

We sailed on, almost halfway through the cruise, collecting more and more samples of seawater and taking the pulse of the Gulf's currents at frequent intervals. The excitement of our preliminary results gave way to the repetitive effort of completing the survey and knowing there would be months of post-cruise data processing needed before any conclusions could be reached.

What tension there was aboard ship radiated from Jim and Vince. It was a tension that drew them to the rail in cycles coinciding with our frequent approaches to the Somali coast, now only a tan smudge on the southern horizon. Their heightened state of alertness was made clear by the number of times they raised their binoculars and by their reduced tolerance for conversation.

Vince had night duty, which earned him the nickname "Shadow." He was more often than not skulking around the dark recesses of the ship, so I shouldn't have been startled one night when he suddenly appeared beside me with an expensive pair of night-vision glasses hanging around his neck. He held his ever-present cup of black coffee, which seemed like a permanent extension of his hand. Vince took up his perch and began to scan the horizon for other vessels.

"How are things going?" I asked him.

"OK."

"Seen anything?"

"No."

"What about that ship over there?" I said, pointing to a series of lights off in the distance.

"I'm not worried about ships. It's the small fishing boats and sailing dhows that are likely to give us trouble."

"What do you think our odds are of getting mixed up with something?"

"Fifty-fifty," he said without lowering his binoculars.

I retired to my cabin feeling a good deal less secure than I had a few minutes before.

Amy was lying on the top bunk listening to an audio book on her tape player. Her feet, clad in white socks, hung over the side rail. She looked so relaxed I decided not to tell her about Vince's 50-50 prediction. I lay awake thinking about the statistics. In the end, I decided the number probably only represented Vince and Jim's desire to see some action.

The next morning there was an unusually high level of activity on deck. Jim had scheduled even more elaborate training exercises. Two crewmen prepared to launch a Zodiac rigid-bottom inflatable tender over the side while Vince, Jim, and the chief mate adjusted bright orange life jackets around their chests.

The operation seemed simple enough; the crew was to use

fire hoses to prevent the attackers, Vince and Jim, from coming aboard. As the first of many mock boarding attempts began, two deckhands dragged the heavy, stiff hose from one side of the ship to the other, while a third did his best to keep the jet of water directed at the Zodiac. From my vantage point above the commotion they looked like a bunch of middle-aged men wrestling a 40-foot greased and vomiting python.

Vince tossed a metal hook over *Knorr*'s railing and tried to climb the attached rope ladder. More than once he took a blast to the head and the sound of high-pressure water slapping against flesh preceded his fall into the water. There were real hazards for Vince in this exercise, such as falling back into the small boat or worse, falling too close to the outboard motor's propeller.

For 30 minutes, amidst much water and a spaghetti-like expanse of hose on deck, the crew prevailed and kept Vince off *Knorr*. It seemed that as hard as it was to maneuver a gushing fire hose, it was harder to be on the receiving end of all that water while trying to climb the side of a ship.

When the tender was hoisted back on deck Vince hopped out first, his pale skin covered in red welts. Jim swung two deeply tanned legs over the gunwale and I note yet another contrast between the two men. If Vince was the quiet chameleon, silently blending into his surroundings, Jim was walking neon. Something of a showman, Jim was at his best in front of an audience. He stood there without saying a word. The crew gravitated toward him, forming a loose ring around the man, anticipating his critique like students awaiting their grades. Vince had disappeared, but his wet footprints lead below, probably back to his cabin where he could dry off, take a fistful of Tylenol, and sleep.

Two days later, it was with a promise of more of Jim's showmanship that I extricated Amy from her computer and walked with her out to the aft deck. After the anti-boarding drill, Jim announced that he had arranged for additional training for anyone who would like to participate. "We don't have to do it. We'll just watch," I told Amy as we climbed steep metal steps to the main level.

"Why can't you watch by yourself?" Amy shielded her sensitive eyes as we popped out into the bright sunlight.

"Think of this as a date. I even made us something to eat." I handed her a hot bag of microwave popcorn and pulled two sun-bleached green plastic chairs into the only shaded spot available.

Knorr's fantail was lined with an assortment of ship's crew and members of the science party. To a man every one of them was, well, a man. Though there were a number of large breasts, none had lactating potential. The most prominent pair belonged to a morbidly obese research assistant whose best weapon was his ability to crush anyone he sat on. The remaining 10 commandos-in-training were a cross-section of the ship's male population, all looking eager to begin something rarely, if ever, done on a research vessel.

The aft deck could be configured to do almost any type of work at sea, from trawling for fish to deploying heavy moorings. I imagined that it was this kind of versatility that Jim wished to create in the men standing before him: men of science by day, instruments of death by night. In reality he was much more pragmatic.

Jim's voice boomed out. "What I hope to do in the next couple of weeks is to teach you a few simple techniques that will allow you to defend yourself and the ship when all evasive tactics have failed, or when confronting stowaways who may be violent."

"So why don't you join them?" Amy asked me.

"I subscribe to the 'I don't have to run faster than the bear, just faster than you' form of self-defense," I said. "Anyway I suspect this training might be more dangerous than anything we encounter out there." I pointed over my shoulder in the general direction of Somalia. I'm not sure I believed the last part of my own statement, but it provided me with an excuse for not having to throw punches at Jim again.

The next hour was filled with a lot of yelling and bruising and a token amount of bleeding. Amy had long since closed her eyes and become immersed in the words of a faceless narrator streaming through her headphones, oblivious of the stray drops of sweat that occasionally flew in our direction.

Just before I, too, had enough of the entertainment, Jim called everyone's attention to a white plastic knife he'd taken from the galley.

"Today we are going to practice subduing an attacker who has a weapon. Jerry, I want you to lunge at me with this knife. Don't hold back, really try to stick me," he said with great confidence.

Jerry "Boats," the boatswain, was a tall, lanky guy who, though not exceptionally muscular, had a significant reach advantage over the stocky Jim. Boats took the knife and thrust down from his six-foot-two-inch frame at all five-foot-nine of Jim. In an instant there was a bone-crunching thud as Boats' hip hit the steel deck grate and the knife appeared in Jim's hand.

"OK, now pair up and do it just like I showed you." Jim handed out more toy weapons.

I closed my eyes while the sun penetrated my skin like a deep tissue massage, and forced my body to relax despite a gnawing doubt that if pirates came aboard they would be carrying *plastic* knives.

The next morning a diffuse grey stain spilled over the horizon where the sun should have been. *Knorr* was enveloped in fog. Fog was the last weather phenomenon I expected to encounter near the equator, but after endless days of the same sun, the same ocean, and the same endless haze, it was a welcome change. It also was excellent cover to shield us from potentially dangerous Somali boats, craft unlikely to be equipped with RADAR.

Cloaked in her hijab of mist, and less than 20 miles off the northwest coast of Somalia, *Knorr* slowly proceeded south, toward land. Amy was bent over the chart table, struggling to read depth contours with a magnifying glass.

"Twelve hundred meters," I called out, reciting depth measurements from the digital fathometer. Amy's jaw tightened with every set of numbers I read. She was counting on *Knorr* reaching the desired depth prior to getting closer than 12 miles from the coast.

At 12 miles out the water was much deeper than the chart indicated. "I think this is as close as the captain wants to get," Amy said to me in a hushed tone.

The State Department had been recommending that all U.S. vessels stay at least 50 to 100 miles from the Somalia coast. As this was unreasonable for our purposes the captain had maintained a play-it-by-ear attitude, letting the situation dictate our limits.

"Why don't you go to his cabin and ask him if it's OK to go in further?" I asked.

"He'll probably say stop here. The current I'm looking for is in closer." Amy paused for a moment, tapping a pencil thoughtfully against her cheek. "The third mate is on the bridge. I think I'll ask him to keep going and see what he says."

I could not believe my ears. This from my wife, the rule-follower, the one who refused to take the tags off furniture cushions for fear of going to jail. For the first time since I'd known her, Amy was actually trying to get away with something.

On the one hand I was of proud of her; her science was important enough that she was willing to bend the rules to get it done. But my thoughts kept going back to Vince's 50-50 prediction. Maybe I should have mentioned it before. Did she have to get daring now of all times?

"Bridge, this is the lab," Amy called into the intercom.

"Go ahead lab," chirped the third mate, a cocky 20-something fresh from the maritime academy.

"We need to get a little closer. We're looking for 900 meters," Amy said with forced flatness.

"No problem, just let me know when you want me to stop."

Amy looked up at me with the face of a 4-year-old who'd just reached the cookie jar for the first time. Then she looked down at the chart, still smiling.

Ten miles from shore, and the water remained much deeper than the chart indicated. "I guess we keep going," she said to me.

There were two things, I reminded myself, that should keep us out of any trouble. The first was our protective shroud of fog. The second was that the north coast of Somalia, unlike the eastern shore, was a fairly desolate place. The chance that anything of consequence was near us, on land or at sea, was remote. I began to relax.

Knorr slowly drifted closer to land. Nine miles, seven; at six miles we were beginning to wonder if the shallow water existed at all. At four miles the captain burst into the lab with the look of a man who had just swallowed a lit cigarette. "Do you know where we are?" he yelled, looking straight at Amy.

"We've been following the fathometer," Amy replied, without letting any air escape her lungs.

"I suggest you look outside," the captain said, with lots of air escaping his.

Everyone in the lab paraded out the nearest door. The captain headed up the stairs to the bridge, taking the steps two at a time. Jim's distressed voice boomed over the PA system and rang off the metal walls: "Condition alpha, condition alpha."

Walking out on deck was like stepping onto a movie set. Any trace of fog had evaporated faster than our story about the fathometer.

"What do you see?" Amy asked me.

"This looks like a good place to stop. I see a town, a fleet of large fishing boats, a road on the coast with trucks moving along it, and if I squint I think I can make out people."

Some of the color drained from Amy's face. Whether it was from embarrassment at being chastised by the captain or from uneasiness at our situation wasn't clear.

Suddenly I caught the earthy smell of land. It confirmed that we were much closer than we should have been.

All around me crew were rushing to what could be called battle stations. Fire hoses were charged with water, ready to repel boarders, and lookouts positioned themselves at each end of the ship. I glanced up to the bridge wing and saw Jim scanning the harbor, his eyes squinting into the haze. He shifted his weight from foot to foot like a man trying to hold in a barium enema. He was not at all happy about the situation, but fortunately for Amy, it was too early to assign blame.

The captain appeared beside us. "I'm going to let you finish this station, so make it a quick one," he said sternly, but his voice lacked any anger.

Amy was so surprised that she could only squeak out a short "thanks."

Over the next 30 minutes the rosette was lowered, brought back up, and the necessary samples taken. With each water sample bottle filled I looked up from my work to scan the shore. There was little talking among those of us on deck. It seemed that everyone else, like the captain, wanted to get the cast over as quickly as possible.

When we were done we headed north away from the coast without incident. Things had worked out this time, though I strongly suspected that Amy would go back to refusing to eat grapes out of the shopping basket until they were paid for. As for the third mate, I hoped he could still sing bass notes after his meeting with the captain. For me it was also a chance to see a side of my wife that I had not seen before, that of someone who was willing to take risks and push the edge. Living with declining vision was a lot like stepping into the unknown, and she would need all the confidence she possessed to keep taking those steps. I had all the confidence in the world that she could do it. I only hoped that she did too.

Our little sojourn would not soon be forgotten, and though the name is not found on any official chart, this part of Somalia will forever be known to the members of this cruise as Amy's Beach.

14

THE ALARM CLOCK HAD BUZZED TO LIFE like so many pissed-off bees at 2300 hours. Three hours before, I had finally coaxed my body to sleep. I was at the end of my 18th day of work on the night shift. I called it the night shift, but the other watch was technically subject to just as much darkness, and in theory the division of labor was equitable. I did not share this sense of equality because my circadian rhythm couldn't reconcile being awake at a time when my body felt the most tired. The constant battle of trying to force sleep in daylight and waking up in the dark was one I felt destined to lose. A midnight breakfast of reheated linguine in clam sauce from that evening's dinner was made worse by my having to look at it through half-closed eyes.

On the chart table in the lab someone had knocked over the pink pig. I leaned over the chart and stood him back up over what approximated our correct location. The little pig had become something of a mascot. He was tiny, about the size of my thumbnail. A cast-off from some game, long since gone. We used the pig to represent *Knorr*'s position along the lightly drawn course line. For me the only thing worse than coming on watch and seeing the pig at the beginning of a long line of closely spaced stations was coming on watch and seeing it at the beginning of a long transit, which meant I could have slept three more hours. If he was as tired as I was I could not blame the pig for wanting to lie down on the job. The midnight-to-noon watch that Amy had volunteered us for was taking a toll on my ability to stay alert too.

I was not the only one on my watch who found this schedule confusing. That evening, staggering into the mess deck, I found Amy alone at a table, leaning on one elbow. "So this schedule is taking its toll on you too?" I asked, sitting down across from her.

"What do you mean?" she said. Then, realizing that she had a spoonful of Raisin Bran in milk in one hand and an open bag of BBQ potato chips next to the bowl, she added, "So maybe I'm a little off."

"This is just what this article I'm reading is about," I said, opening a copy of the magazine *Professional Mariner,* which I'd found in the ship's library.

"If you insist on reading at the table, you'll have to do it out loud so I can listen too," my wife responded.

"I was planning on it," I said, though the thought hadn't really crossed my mind. I pick up one of Amy's potato chips.

"So what's this article?" she asked.

I turned the pages, leaving orange, greasy fingerprints on the corner of each. "The title is 'Asleep at the Wheel.' It goes into great detail about how odd watch schedules aboard ships can lead to accidents as a result of insufficient rest."

Amy laughed. "Do you consider yourself a hazard?"

"Only to my roommate, if I continue to lose sleep."

She swallowed a mouthful of cereal.

I read on, hoping to make my case for a better watch schedule. "The article describes a test the Coast Guard uses to determine if sleep deprivation was a factor when they investigate a maritime incident. The test is a numerical index determined by multiplying the number of symptoms of fatigue such as forgetfulness, desire to sit, and muscle aches by a factor of 21.4. Added to this is the number of hours since waking. Multiply the result by 6.1 and subtract the number of hours of sleep in the last 24 hours. Multiply this result by 4.5 and the answer is the level of incapacitation due to fatigue." I ate more of her chips, the spicy flavor assuring me that at least my taste buds were awake.

"The Coast Guard really uses this?" Amy moved the chip bag out of my reach.

"I'm skimming, so I may have some of the numbers wrong, but the concept is valid," I said.

"You took this quiz? What did you get for a number?"

"It says here anything over 50 is hazardous. I'm guessing Captain Hazelwood, when he tried to drive the Exxon Valdez down the Alaskan Interstate, probably rang the bell at a hundred. According to my addition. . . ." I tallied my results on a napkin, ". . . I will need to sleep until June just to get under 500."

"You'd better check your math," Amy said with no sign of sympathy. "In any case, I'm not changing the watch schedule just for you. Besides, I can't; it would impact everyone else who doesn't have a problem with it." She cleaned up the remains of her meal.

I didn't expect her to change it, but I had thought she would be more sympathetic. I walked back into the lab. The pig was wallowing between stations so I had some time to kill before starting work. Having survived our detour into downtown Somalia, we were heading back north for another transect of the Gulf. This time the captain had made it clear that 12 miles was as close as *Knorr* would sail to either the Yemeni or the Somali coast. The ship plodded along the north-south lines, sailing for an hour and stopping on station for two hours. The one-hour steam was just long enough for us to log the data from the previous cast, set up everything for the next one, and find some food in the galley. Every time we reached the end of one line and moved east again, it took anywhere from three to four hours to get to the top of the next north-south line.

Someone from the off-going watch said that the bridge reported a lot of bioluminescence in the water, so Amy and I took a walk aft on deck to check it out.

Standing at the stern, I looked down at the wave pushed outward from *Knorr*'s hull as the ship glided along at 10 knots, cutting through mile after mile of Gulf water. The wind was starting to pick up ever so slightly from the zephyrs of air we'd had earlier in the day. A sliver of a moon sat high above our heads, like one of those curved knives favored by the people of this region, and in its faint light wavelets could be seen cresting on the sea surface. At that moment our world consisted of three sources of illumination: the reflected white moon, the yellow incandescence spilling out of *Knorr*'s portholes, and the electric blue of bioluminescence churned up behind the ship.

Within the widening V of *Knorr*'s wake there was a subtle glow. At first it was hidden by the saturating reflection of the moon, but as the moon set and the sky darkened, it was obvious that the neon light was originating from within, not above the water's surface. It grew in intensity until it seemed the ship was sending out sparks from her thrusters. I was hypnotized by the sight and wanted to know if Amy could see it too.

"Do you see anything?"

She paused, squinting into the darkness. "Maybe a little. I miss it a lot. I remember those little things about being at sea."

I tried to tell her what I was seeing, before realizing that she had a better picture in her memory than I could describe. One of the worst things about being blind is having to live certain events through the filter of other people. As a guide I had come to this realization only recently, though it didn't make the job of describing bioluminescence any easier.

It was those millions of little lights trailing behind *Knorr* that reminded me that the water we rode on was not as devoid of life as it first appeared. The cycle of procreation, growth, and death was constantly in motion below us, unseen except for the occasional signals that broke the surface.

One of my most memorable nights at sea was in conditions similar to this. A few years ago, at Amy's suggestion, I had found a position with the Sea Education Association on a sail-training and oceanographic research schooner in the Caribbean. I was alone at the bow one night as the ship undulated over a gentle swell. There was no moon and the only way to find the horizon was by the abrupt end of stars. Suddenly, from under the bulwarks, there was an explosive rush of air, as if someone below me had just finished a deep dive on a single breath and was now frantically breaking the surface. Another whoosh of air, the same as the first, then another and another until it felt as if I was in a factory of steam engines. I knew it was dolphins, but what appeared were not dolphins, but the painted outlines of dolphins. Long, wispy brush strokes of blue trailed one behind the other like contrails as they disturbed the bioluminescent algae around them. When they darted in and around each another the streaks of blue intertwined in an electric helix. That night I had watched until the last dolphin left the bow wave, the sapphire trails fading to black.

"Have I ever thanked you for getting me out here?" I moved closer to Amy at the rail.

"A little while ago you didn't sound too happy," she said, still straining to see anything light up in the water.

"I should probably stop whining about lack of sleep. If I were in bed I'd miss all this. Anyway, thanks." I put my arms around her. For once Amy didn't seem to care if there was anyone on deck to see us.

"You're welcome. I think I got a pretty good deal too. You've turned out to be a reasonably decent guide."

"Well, the trip is not over yet, so don't be so quick to give me a passing grade," I said as we turned to go back down below.

Amy walked back to her lab desk, her eyes now having to adjust to the glare of artificial florescent lights. I searched out a book in the ship's library that would explain the light show we'd just witnessed.

As I expected, the books in *Knorr*'s library had plenty of references to bioluminescence. The sparkling neon in the water the ship passed through was probably the result of large colonies of dinoflagellates, single-celled algae. Some species of dinoflagellates were best known around coastal communities as toxic red tide blooms. I chuckled to myself at the name dinoflagellate; what came to mind was the image of a brontosaurus on a holy pilgrimage, beating himself with a leather whip. In reality these algae are so tiny that there may be thousands of them drifting in each gallon of seawater. They have no destination and drift at the whim of the currents. Scientists cannot say what possesses these algae to make such a grand announcement of their presence. In the case of bioluminescent algae, the process is chemical; when the algae are jostled by the water, stirred up by the wind or by the bow of a ship for example, the enzyme luciferase catalyzes the oxidation of the chemical luciferin. This oxidation releases energy in the form of blue light.

Knowing the chemical origin of bioluminescence did not diminish its splendor for me, but only made it more magical. It seemed ironic to me that this light from the deep, a glow of such spiritual magnitude, should be named for the devil. But as I thought of it, it seemed no less ironic than nature stealing the sight from a woman whose passion sprang from her sense of observation.

I heard *Knorr*'s engines throttle back and felt the resistance of the sea slow the big ship down. It would take about 10 minutes for the mate to position *Knorr* where Amy wanted us. I closed the book and went to the lab to make preparations for the next cast. Inside, Amy was talking with one of the other scientists. They stood in front of her video magnifier, which displayed an enlarged satellite image of the surface circulation of the Gulf. Amy was smiling and the pig sat resolute upon our position on the chart.

15

It would be two hours before we arrived at the next station and I took the opportunity to check out the leftovers in the galley fridge. Entering the mess deck, I found Steve bent at the waist, his head lost in the open mouth of a stainless steel refrigerator.

"So what looks good?" I asked, assuming there was a head attached to the torso.

"Let's see, salmon steaks with dill, pork chops and vinegar peppers, pork fried rice, or very rare roast beef with horseradish sauce. Take your pick."

It was 4:30 in the morning. None of it sounded the least bit appetizing. So I found my old standby: toasted bread with Nutella hazelnut chocolate spread. After some minutes, a loud bell in the microwave oven chimed and Steve pulled out a steaming plate. It smelled like a mixture of everything he'd just listed. He sat down across from me and began strip-mining the mountain of food in front of him. "What's new?" he asked.

The question was rhetorical; we were living so closely together that if something truly new and interesting were to happen it would be common knowledge within three turns of the propellers.

"Nothing," I said absently.

"Just what I want to hear, no problems," he said between big forkfuls of a now unidentifiable blend of animal, mineral, and vegetable.

"I once read that going to sea has all the benefits of suicide without any of the inconveniences," I said, smearing a sticky gob of Nutella on my toast. "The mind-numbing effect of routine."

"Believe me, it's a little different when you're the scientist in charge," he responded. "Though I can remember the days when there was no requirement for me to have an original thought. Everything preplanned; when to sleep, when to eat, what was for dinner. I could do four weeks on autopilot."

"Prison without the orange jumpsuits," I said.

"I never saw it like that. Anyway, it's different now. Satellite phones and email changed everything. Now my wife can send a message asking when the dog last got his shots, or where did I put the hedge trimmers?" Steve gazed wistfully into what looked like mashed potatoes.

"Do you and your wife ever sail together?"

"Sometimes, but when you're in the same business it can be a power thing. Each of us has our own ideas about how stuff should get done."

"The captain can marry you, but he can't grant a divorce," I said, anticipating his thought.

"Yeah, and there are no lawyers at sea, though I could use one now. I'm having a little 'division of labor' matter with some people in my watch." He rummaged around the condiment box on the table, hunting for a particular hot sauce.

Thankfully, from my perspective, life aboard ship was orderly, with an emphasis on basic comforts and safety. It was an enclosed environment where other people were responsible for my well-being and nutrition. All that was expected of me was that I learned new things and behaved myself. Maybe this was not as much like prison as it was like day care.

Back at my computer, I checked my email and found that my conversation with Steve was prophetic. The first of many messages contained word from our house sitter that one of our aquarium fish has died. I broke the tragic news to Amy.

"Was it one of the orange ones or the white one?" she asked without looking up from the screen.

"She didn't say. Probably couldn't even tell. They all turn a fleshy shade of white after floating upside down for a day or two."

Since we couldn't be there for a proper sendoff, I decided to conduct our own ceremony aboard ship. Crammed into the head, solemnly standing over the porcelain bowl with the black institutional seat, the kind with the front cut out, I proceed to flush a proxy paper fish down the john. It turned out to be quite a spectacle. Since seawater was used to flush the toilets, the same algae that had caused the ship's wake to glow blue the other night also irradiated the bowl with each flush. The ability to observe one of nature's most beautiful biochemical processes in my own toilet was inspiring.

To me this death of an aquarium fish highlighted our isolation. What would happen if someone higher on the food chain passed away? Just about anywhere on the planet, if you get a call that a loved one has died, you can hop a cab to the nearest airport and be at a funeral home halfway around the world before the corpse is cold. At sea, redirecting a research ship to the closest port can take days and cost tens of thousands of dollars.

I queried Danny, the third mate, about this.

"It depends on the circumstances," he said. "How far away from a port we are, your relationship to the deceased, how strongly you feel the need to be there, are all things we need to consider." He looked up from the chart he'd been plotting on. "If your dog dies you get to send a fax home. If it's your mother, we airlift you off the deck. It's all up to the chief scientist."

There wasn't much time to reflect on the philosophical differences between mourning my dog and mourning my mother, because I was interrupted in my thoughts by Francis, the AB. "You know what day tomorrow is, don't you?" he asked. His bushy black eyebrows stabbed upward the way a silent movie villain's would to convey sinister intent.

"Thursday," I said, knowing immediately that this was not the right answer.

"Ha, the bliss of the ignorant," he chirped out in a voice two octaves higher than his stocky 240-pound frame made plausible.

He shot me one last devilish glance and walked to the other side of the bridge, chuckling to himself. As his laughter faded away, my reflection in the window said to me, "If this ship goes down, do not get into that man's life raft."

Danny offered an explanation before I could ask.

"Equator crossing."

"Oh," I said, with a sudden understanding of Francis' strange behavior.

Knorr had actually crossed the equator earlier in the trip but because we were only one day into the cruise, the age-old tradition of hazing first-time crossers, called pollywogs, and turning them into Shellbacks, had been postponed. Additionally, a break in the science schedule afforded the crew a leisurely, unhurried time frame during which to plan an elaborate ceremony.

The line-crossing ritual goes back hundreds of years and is probably the world's second-oldest professional fraternity. The first-oldest, also related to sailors, is based in most ports of call. These hazings, like most initiations, were probably created out of the need to form close bonds among shipmates. In the early days of sail there was a greater reliance on other members of the crew when it came to survival, and having a member of the brotherhood, one who has done his sea time, beside you in the rigging led to a greater feeling of security.

The ritual continues today as much out of a sense of entitlement by past initiates as out of tradition. The attitude "If I had to go through it so do you" remains not only aboard ship but in such venerable institutions as legal bar exams and the grueling rotations of medical school. I had seen photos in the scrapbooks of seagoing friends that showed this rite of passage being celebrated by disgusting activities that seemed to differ from cruise to cruise depending on the creativity of the Shellbacks aboard. Judging by the sly grins and strange looks I was getting from veteran sailors passing through the bridge, I feared this lot to be exceptionally creative.

Membership in this sacred society was confirmed with a laminated card, with the ship being divided into the card "haves" and the card "have-nots." Danny confessed to me that he had forgotten to bring his card on this trip. He was in his early twenties, with wire-rim glasses and a wispy, adolescent mustache growing from under his nose. Tussled blond hair radiated out in all directions from under a well-worn baseball cap. "Haven't cut my hair since leaving the Maritime Academy," he told me at the beginning of the cruise. Now, on *Knorr*'s bridge, he related to me a torturous equator crossing endured on his senior-year Academy cruise.

"Can't you just tell them you've been through this before, you know, give them your mother's maiden name or the last four digits of your social security number and have them look something up?" I asked, incredulous that he would be willing to subject himself to it all again.

"Oh, you can try and get someone to vouch for you, but nobody on this boat would do that for me. The few people on board who know I've got a card would like nothing better than to see me go through the process again," Danny said with a smile.

I stared for a moment at the spinning arm of the beam indicator glowing green on the RADAR screen. "What if I just tell them no thanks, I don't want to join your club?" I said, a little apprehensive about the whole thing.

Danny gave me a quizzical look. "I suppose you could. Nobody's putting a gun to your head, but why wouldn't you do it? Do you really want to break a tradition that's as old as ships?"

"No, I guess not."

The next morning all pollywogs were called to muster in the mess deck. I was surprised to see Jim and Vince among this group of 10. They were both wearing defiant smirks. Jim seemed to be looking forward to the challenge. The rest of us were members of the science party with a couple of ship's crew thrown in. Remarkably, Amy was also among the uninitiated.

"I can't believe that with all your sea time you've never crossed the equator," I said to her while we waited for something to happen.

"My work has always been in one hemisphere or the other, so I've always flown across the equator," she said, without elaborating any further. It was hard to tell if she was more concerned over the physical challenge we were about to face or her potential loss of dignity in front of colleagues.

The remainder of *Knorr*'s complement, who only yesterday had seemed like long-lost brothers, stood with crossed arms along the perimeter of the mess deck, their faces expressionless as polished granite.

Steve, Amy's most senior collaborating scientist, stood before us dressed as Davy Jones. Wrapped in a bed sheet, he wore a cotton mop wig dyed yellow, and held a toilet plunger scepter in his right hand. Steve introduced King Neptune, and Francis appeared in a similar costume with the addition of a gold-foil crown atop his balding head.

Steve continued the ceremony by assigning pollywog names to the 10 uninitiated members of the ship's company. Earlier in the cruise I'd made the mistake of staging a shot for my video camera, and my name became "Cecil Wanna B. de Wog." Amy, in the spirit of her assault on the Somali beach, had earned the moniker "Condition Alpha Wog." In case any of us should forget our new identities, cardboard name placards had been prepared for us to wear around our necks.

After the naming rite, King Neptune addressed the assembly. "Wogs shall wear all underclothing on the outside," bellowed Francis in his best impression of a regal voice. A groan from my fellow pollywogs seemed to stimulate a last-minute idea in the king. "And all articles of clothing must be turned inside out." Francis smiled, proud of himself over this last twist. A long list of other rules followed before we were released to execute this new fashion statement.

"You know, you don't have to do this," I said to Amy as she dug through her drawer for a more chief-scientist-looking pair of panties.

"If not on this cruise, then it would be another one. Better to get it over with," she replied as she struggled to put a bra over her tee shirt.

A short while later we reported back to the mess deck properly attired. I looked around the room at my nine compatriots. I couldn't believe that my mother had foreseen this day so many years ago. Having clean underwear WAS important. Blindfolds were wrapped tightly around our eyes, and we were led like a chain gang, each holding the shoulder of the person in front, up to the main deck.

I had the same feeling in the pit of my stomach that I got when a roller coaster chugged to the top of the first big drop. There was murmuring among my fellow wogs as our human chain serpentined through passages. Every few steps my shin hit the lower lip of a watertight doorway or some other obstruction. I could really sympathize with Amy, as my hands tightly gripped her shoulders, who must negotiate these obstacles every day. Passing through one last door I felt the sun on my face, indicating that we had finally reached the main deck.

Card-carrying shipmates had formed a gauntlet on either side of this ridiculous conga line. As we passed by I heard their jeers and felt viscous liquid drops hit my head. Some of them began to run down my cheek and into my mouth. It was the sweet taste of Maple syrup; thankfully, my first thought that it was bird shit was unfounded. After a few steps, ketchup, Tabasco sauce, and oatmeal became the gunk of choice. Before I could fully appreciate what we must look like to our tormentors, a deluge of cold water from the ship's fire hose rained down on us. I should have been grateful for the shower, but all the water seemed to do was bind the various elements together into a putrid paste.

On deck, the salt water evaporating from my head left behind a growing collection of crystals that began to itch my scalp. Our pollywog group smelled like the bottom of a restaurant dumpster as we stood together, fermenting in the hot sun.

"That's it for today, boys and girls. You can take the blindfolds off," proclaimed King Neptune. "Muster in the galley at 0700 tomorrow for further instructions." There was a pause and a wag of the royal finger. "And remember, all of the rules about dress are still in effect."

Back in my cabin, under a real shower I managed to remove enough flour, egg, and jelly from my ears to make a respectable French pastry. I realized that playing with food like this is every 3-year-old's dream day, but at 40 years of age I should not have to be picking oatmeal out of my belly button.

The next morning proved that the first day was only an appetizer, in every sense of the word. Still clad in our inside-out attire, Amy and I joined the rest of the initiates for breakfast. The mess deck was much more crowded than usual, and one look at the menu board told me why. A special meal awaited us and an exceptionally large audience was present to watch us eat it.

Amy and I shared a table with two of our tortured compatriots. The rest of the ship's company crowded the doorway for a better view. The steward placed a covered dish before each of us and lifted the lids off with a flourish. On my plate were three mounds of a gelatinous substance, each a different neon color. A fourth item that looked like a calcified cat turd sat ominously near the edge of the dish.

"Eat up, Wogs," cried King Neptune, his booming voice jigging the frayed mop yarns dangling along his cheeks.

"This is one time where not being able to see well has its advantages," Amy whispered to me.

I closed my eyes in hopes of mimicking her condition and picked up a spoonful of glowing goo. I guided it to my mouth. Before it got between my lips I heard the woman next to me gag. It took two more tries before I found the courage to put the stuff in my mouth. The taste was not so bad, but the texture was unnerving. After a few mouthfuls I attempted a very child-like maneuver and tried to make what was left disappear by rearranging it on my plate.

"Finish everything or you'll get seconds," yelled the steward from behind his stove.

With no options left I choked down what remained in front of me, and in an attempt at chivalry, took a few bites off of Amy's dish.

After surviving breakfast, we were divided into work details and sent throughout the ship with a small group of overseers. For the next three hours Amy and I scrubbed the foredeck with tiny brushes, while enduring more saltwater hosings if we appeared to our overseer to be the least bit laggardly.

At noon, during what was supposed to be a break for lunch, we found ourselves deposited in the ship's library, reunited with our fellow initiates. Everyone was anxiously comparing notes and talking about the most disgusting thing they'd been required to clean. Galley duty won out easily and appetites for lunch had diminished considerably after just a few minutes.

Before we could complain much more, the Royal Scribe appeared at the library door with an edict. "In two hours you will muster on the fantail to entertain the royal court. If you don't want to incur the king's wrath I suggest that you create something funny." Since this job seemed far cleaner and less odiferous than anything we had been required to do thus far, we happily set to work.

With props in hand and a couple of hours behind us for writing and memorizing our parts, the 10 of us assembled on a makeshift stage on *Knorr*'s fantail. The king's court in their best janitorial finery sat front row center, while others were sitting in chairs or standing at the rail on an upper deck. The show began, and I immediately saw that the skits we were performing were only an excuse to distract us from the onslaught of more condiments.

Encrusted in both edible and non-edible materials, I thought things could not possibly get any more uncomfortable. I was grossly mistaken. The time of judgment was at hand. All pollywogs were rounded up and locked in the wet lab, a small room immediately off the main deck. Each pollywog was segregated from the rest and brought before the royal court to hear the charges against him. Cries of protest filtered into the confined lab space and those of us inside glanced around nervously, wondering who was going to be next.

My turn came; I was dragged out into the bright sunlight and seated before the king. I had trouble listening to the charges being read against me. I was preoccupied with a numbness in my butt caused by the chair I was sitting on. It was made out of solid ice. Any composure I tried to regain was lost when a look between my legs revealed the tentacles of a half-frozen octopus reaching up through my seat.

Before I knew it the court has passed judgment, and if there was an oceanic parallel to a kangaroo I would use it here. What followed next was a blur of repulsive acts. Two of the larger crew members lifted me off my seat. Since my ice chair had long since frozen to my shorts, it lifted up with me. As second later the block of ice went clattering back down on the deck, leaving a section of pink tentacle dangling from my butt. My penance, it seemed, was to crawl through a tube named "the whale's asshole." Bending down, I stuck my head in a tarp-covered basin about three feet wide by 20 feet long and understood how the thing got its name. The stench that rose from the opening made me want to go back and sit on the octopus. But from the looks of the people around me, this was a one-way tunnel.

With a deep inhalation I crawled on hands and knees into the opening. From behind me came the words "On your belly, Wog, or you have to go through again." I dropped down into four inches of brown, slimy ooze. It looked like vomit and I tried my best not to add any of the real stuff to the mix. Chunks of solids found their way into my pants and shirt, and by the time I emerged at the end, my hair was plastered with thick sludge. I was brought again before the king.

"Recite the act of contrition to His Majesty or go through again," Steve said, waving his toilette plunger in my direction.

I had a vague recollection of being given a slip of paper yesterday with something written on it. But before I could even offer an incorrect attempt, Steve added, "Unless you would rather kiss the baby's belly."

I looked over at a figure sitting next to the king. The "baby"' was a potbellied, six-foot man dressed in nothing but a diaper and the familiar mop hair. In his navel was a large black olive surrounded by something that almost, but not quite, resembled peanut butter.

I had a sense that eating the olive out of this guy's hairy navel was inevitable, so to minimize my transit through the cesspool, I consented. I closed my eyes and bobbed for the black pearl.

Nothing remained but the simple task of kissing a dead fish head. On my knees, and looking at the empty space where the olive had rested a few minutes before, I found that the thought of kissing the severed head of a huge Mahi-Mahi was anticlimactic. Pucker up fish!

My ordeal now over, I stood back and watched as Amy was led before the court. Steve began reciting her charges. "Condition Alpha Wog. You have been found guilty of the following offenses. . . ."

Amy was not listening any more than I had been; she was too preoccupied with trying to find out why her chair was so cold. At least she couldn't see the octopus.

When at last Amy came sputtering out the end of the tunnel, I saw a vision of how I must have looked only a few minutes before: monochrome brown from head to toe. When they sat her down and asked her to recite the act of contrition, Amy, to everyone's surprise but mine, rattled it off without a mistake.

If any deference was paid to Amy because of her rank, sex, or both, certainly none was given to Jim. If anything his treatment was more mean-spirited. Members of the crew manhandled a struggling Jim before the court. When his charges were read he dropped his pants and waved his bare butt in the direction of the king. With this act he sealed his fate. Three men wrestled him to the ground, placed a blindfold over his eyes, bound his hands, and proceeded to lift him in a harness off the deck with the ship's crane. Like a birthday piñata, Jim was swatted with water from a hose until everyone on deck was wet from the spray shedding off his spinning body.

Shellbacks we had become! The two-day, belated equator-crossing ceremony was finally over and all 10 of us had survived the ordeal. We congratulated each other and walked among the rest of the ship's crew with the confidence of equals. In truth, my shell may not have gotten stronger, but my stomach did, having been conditioned to tolerate the foulest concoctions assembled at sea. I stood in the ship's laundry pulling damp clothes out of the washer. Even after two washings, there was still a stench that would not leave my clothes—an odor so strong that it overpowered the ship smell they'd previously absorbed. Without ceremony, I tossed yesterday's pollywog skin in the trash.

I realized that our isolation at sea required that we create a diversion from the seemingly endless horizon that defined our world. We did this by the bonds we formed with shipmates and by maintaining the traditions and ceremonies of generations of sailors. These were the small comforts that made a long separation from land more palatable. But the ocean is not a static place, and like the transformation of pollywogs into shellbacks, people grow into their roles, as I felt I had. For the first time I sensed that I wasn't just a passenger, but a contributor. I knew that in the future, when we became too old to stand on a pitching deck at sea, Amy and I would remember our shared experiences every time we came across the faded, dog-eared, equator-crossing cards in our wallets.

16

THE LAST OF FIVE INSTRUMENTS was tied down to the brick-colored metal deck of the main lab. The device consisted of two aluminum cylinders stacked end to end, and was suspended in a seven-foot stainless steel frame. The first cylinder was 10 inches in diameter and housed the electronics and battery pack. The second cylinder, slightly larger in diameter and painted bright yellow, was open at one end. I looked through the open end and it first appeared to be empty. On closer examination I could discern the diaphragm. This diaphragm worked like a speaker cone, allowing the hollow tube to act as a resonator, much like an enclosure for a home stereo speaker. And speakers were exactly what they were, though these sound beacons had been designed to work in the deep ocean.

Apprehensively, I connected my laptop to a plug dangling from one end of the sound beacon. Each underwater speaker was tested prior to being shipped to Africa, and I had tested them all again a dozen times since coming on board. These instruments were a key component of Amy's work and I didn't have enough confidence in them, or myself, to leave things alone. I felt like the parent of a sleeping newborn, poking her repeatedly to see if she was still breathing. Amy had given me the responsibility of getting these things ready for deployment, and I didn't want to fail at the task. She had told me stories of instruments released over the side when someone forgot to turn them on, leaving thousands of dollars of electronics to drift uselessly around the ocean. I didn't want future stories like that to be told under my name, so I tested each one again and again.

My concern was not without some justification. Upon opening the third of five shipping crates, I lifted the wooden cover off to reveal sound beacon number three lying in a pool of its own oil. The cause of death was determined to be a torn rubber bladder in the resonator tube. The leaking mineral oil was necessary to fill the compressible air space inside the unit. Without the oil the housing would have been crushed under the pressure of the mile-deep water column we hoped to deploy it in. If you fill a balloon with air and take it to the bottom of a swimming pool, the balloon will get smaller. Fill the same balloon with a noncompressible fluid, and the balloon's size will not change. Engineers designing ocean instruments typically use mineral oil because it is electrically inert and will not damage energized electronic components.

"So this is it, the last one. After we get it moored we're done," Amy said, entering the space where I was assembling the sound beacons. "Any problems yet?" The area was littered with empty wooden crates, corrugated shipping boxes, foam packing material from everything we had deployed so far, and the remains of our deceased friend, which would be shipped back for repair, to be deployed on a later cruise.

I couldn't help but think Amy was there to check up on me, concerned that I'd missed something. I wouldn't have faulted her for it. This was her work. No one would blame her for wanting to oversee every little detail. There was too much time and money riding on the performance of these instruments, and she was ultimately responsible for the program's success or failure. Not being able to see enough to verify that the people working for her were getting it right must have been frustrating, but if Amy had that frustration she didn't show it. She seemed much more relaxed than she had at the beginning of the cruise, which I hoped also reflected an increased confidence in my ability.

"This one looks fine. I was just hooking up the computer to check the sound schedule one last time," I said to her as I plugged a test cable into the side of the unit. The laptop interface allowed me to review the internal clock timing, reprogram the sound event schedule, and toy with a myriad of other features; all in all, way too many opportunities for me to screw something up.

This was, by far, much more responsibility than I had had on my first research cruise. If more than one sound beacon did not function the whole float portion of the study could be in jeopardy, and I was sure Amy would find some excuse not to take me to sea with her again. For myself, I had been bitten by the oceanography bug and didn't want to jeopardize my chances of being invited out again.

As I connected the data cable to the back of my laptop I thought about how technology had enabled researchers like Amy to learn more about the circulation of the earth's oceans. Surface drifting floats had long been used to map surface currents like the Gulf Stream. The position of the drifter could be easily determined using radio signals. After months of drifting, the instrument's daily positions could be pieced together in great detail to reveal the long meandering path of the current.

When scientists realized that the flow of water on the ocean's surface did not necessarily extend into the deep water, they faced a significant challenge: how could they measure the direction and speed of a current hundreds or thousands of meters below the surface?

One solution to the problem of seeing the subsurface currents was the use of moored current meters. These meters, with spinning impellers similar to wind vanes, could be anchored to the bottom and configured to float at whatever depth the scientist was interested in studying. The vane pointed into the current and impellers measured the speed of the passing water. After a period of days, months, or years the mooring and all of its stored data would be retrieved.

This method was limited to measuring the current only at the location of the mooring. It was, in essence, like watching cars drive by on Elm Street. Doing so would not yield any insight into the traffic on Main Street. More current moorings would generate a more accurate picture of what was happening, but a lot of interpolation was required to yield even a crude sense of how water was circulating over such a large ocean. To get an idea of what was happening in an area the size of the North Atlantic, many thousands of moored meters would be required, something both physically and fiscally impractical. What was needed was a method of using surface drifting floats under water.

In the early 1970s, researchers developed the SOFAR float. The name SOFAR is an acronym for SOund Fixing And Ranging, and describes a layer of depth within the ocean at which sound moves the slowest. In this layer, which occurs at an average depth of about 800 meters, low-frequency sound waves are funneled horizontally for great distances. Sound created within the channel will stay there.

The float took advantage of this physical phenomenon, just as humpback whales do. This instrument was large; it measured a foot in diameter and weighed almost half a ton. The float could be weighted to be neutrally buoyant at any depth, allowing it to sink and remain at the depth of the current under study. It contained a large bank of batteries powering a speaker capable of producing a foghorn-like noise once a day at a precise time. Since sound propagates very well in water—water being 830 times denser than air—and sound waves within the SOFAR channel are refracted inward, and so do not scatter, the signal from these devices could be heard for thousands of kilometers under water. This characteristic was valuable to the United States: Since the Cold War the U.S. Navy had maintained a line of hydrophones around the Northwest Atlantic called the SOSUS (SOund SUrveillance System) line. These underwater "ears" were used to monitor Russian submarine traffic.

For scientists, it was a simple matter of asking the Navy when it heard the tone from the float at each listening station. Knowing the speed of sound through the water, the position of the receiver, and the time of arrival of the tone, one could get a daily fix of the floats' positions. Connecting the dots of daily position yielded a track of the float moving along with the current.

As the scope of subsurface current research increased, it became apparent that the cost, size, and reliability issues of hundreds of underwater boom boxes were prohibitive. There was also the issue of the navy SOSUS lines being dismantled in the 1990s in favor of more sensitive mobile arrays.

The solution for oceanographers came in the form of the RAFOS float. An astute reader will see that RAFOS is SOFAR spelled backward, and that is exactly what was done to the technology. Instead of sending a signal from the float, new floats were developed with small hydrophones that allowed them to receive sound signals generated by sound sources moored on the seafloor. With the heavy and expensive components now fixed in place, a small, less expensive drifting float, capable of remaining underwater for up to two years, could be deployed. As long as the float was in a position to hear at least two sound beacons, a daily location fix could be gotten. At the end of its missions, a float would drop a ballast weight, rise to the ocean surface, and transmit all of its stored position data to a satellite, which would in turn send the data back to a land-based computer for analysis by the scientist.

For our REDSOX experiment, five sound generators would be placed around the Gulf of Aden. This number ensured that no matter where floats drifted they would be in hearing range of at least two sources.

Amy and I sat quietly, staring down at the large metal tube at our feet. I glance at my watch to check the time and scraped off a piece of dried, brown gunk from the crystal face, another remnant from our equator crossing.

"Fifteen seconds . . . ten . . . five, four, three, two, one," I counted down aloud. As the second hand clicked past 12, a warm sound enveloped the room. The low frequency made it difficult for us to pinpoint its origin, but we both knew where it came from. The tone was pure and unadorned, but still sounded like music in our ears.

It would be at least eight hours before we arrived at the position selected for this mooring. I began to coil up the tangle of cables at my feet, careful not to end up with a knotted mess. This attention to detail was not something I typically practiced at home. There, my collection of extension cords lay in a box, twisted together like mating snakes.

"Can you help me check something on the chart?" Amy asked, once I had finished stowing things away. The large Gulf of Aden chart was spread out on the lab table. "I'm having second thoughts about exactly where this sound source mooring should go. Can you trace out a bold outline of the 1000-meter bathymetry contour? I can't get the big map under the video magnifier," she said.

"What are you looking for?"

"I want to see if there will be any shadows where a seamount or canyon would block the sound."

Amy and the other scientists had studied the sea floor of the Gulf in great detail before deciding where each source would be placed. The bottom of the Gulf of Aden, like most of the seabed, is not a smooth, flat, water-covered prairie, but an undulating layer of the earth's crust. Cliffs, pinnacles, and canyons define a seascape that lies hidden from our view and can only be represented to us by bottom-mapping SONAR. If the water were drained from the Gulf, we'd see something resembling the Badlands of South Dakota. However, unlike wind eroding arid rock, slow-moving currents do little to change the landscape, and the sea floor retains most of its original detail. What change does take place is more spatial, the result of spreading geologic plates.

These bathymetric features meant that the placement of sound generators was critical. A beacon moored in a deep rift would have its signal captured by the confines of the valley. One placed behind an underwater mountain would be unable to project its sound very far, being shadowed by the obstruction. Any RAFOS float traveling in this zone of silence would not be able to record a position until it drifted out of this acoustic shadow. Amy and her colleagues had spent hours back in Woods Hole hunched over SONAR projections of the Gulf's bottom, trying to find five relatively unobscured locations that would ensure an array of overlapping sound signals throughout the region.

I traced the 1000-meter contour in bold ink, and then turned back to take care of one last detail that needed to be completed on the sound beacon.

Each of the five sound generators had been designated by a letter, A through E, corresponding to the mooring locations selected on the chart. I carefully stenciled a large black capital E in indelible marker over the surface of one unit's resonator. I had monogrammed the other units with their letters and signal times so that there would be no confusion when selecting the correct one for deployment. Since this was the last one to be launched the exercise was redundant, but I felt the need to maintain my little ritual.

Having a pen in hand and an object that might never again be seen by human eyes enticed me to inscribe a few personal remarks alongside Mr. E's name tag. It was the same force that motivated people to carve their name in a tree or write poetry on the shell casing of a bomb. Why is it human nature to do such things? I don't have the answer, but I know that NASA spent a lot of money to put a plaque on the *Voyager* spacecraft. Did they really think that some alien creature would be strolling the dunes of a far-off planet, stub the toe of his third leg on *Voyager* and say "Wow"? Probably not, but no matter how slim the odds, humans are fundamentally an optimistic race and that's why I wrote my little note to nobody:

'CAUTION RADIOACTIVE'
If Found Deposit In Any Mailbox
C/O Democratic People's Republic of North Korea
Postage Guaranteed

Despite all I had done, there remained a nagging fear that this unit would settle into the sea floor and remain mute. Walking down the narrow passageway, I realized I was becoming my own flaw in my theory of human optimism. Maybe what I was experiencing was a healthy dose of scientific skepticism. I found myself thinking not about how Amy would react to failure, but about how much data would be lost as a result. Maybe my work with Amy was beginning to teach me how to think like a scientist.

17

THE LOUNGE WAS FAR FORWARD ON *KNORR*. Forward enough that the walls began to take on the shape of the bow, each one curving in and down. Both sides of the room were lined with row upon row of videotapes like black plastic wallpaper. The room was illuminated in the electric blue light of the 32-inch TV, while half a dozen off-watch crew and scientists were slumped in comfortable-looking chairs. I found an empty seat and prepared to be deafened by Arnold Schwarzenegger blasting his way through yet another dense jungle. Arnold seemed to be a favorite at sea, probably because the loud sound track easily masked the noise of the ship. The sound system vibrated my chair, but this far forward there were no cabin spaces nearby, and no one to complain. It was only during a lull in the action that the sound of the bow plowing into the waves could be heard through the steel hull.

Like clockwork, every 10 minutes someone would stick his head into the lounge and ask, "What's the movie?" Without looking away from the screen, one of the slouched heads would answer, and the inquisitor would inevitably say, "I've seen it" and walk off.

Early the next morning we were close to Station "E" where the first sound source would be deployed. The mooring procedure did not look complex on paper, but in practice it required the skill of a small team of technicians and riggers. After I activated the unit and deemed it ready, this group would take over, seeing to it that the instrument was placed safely on the ocean bottom in the precise location Amy had specified.

There was more to sinking a mooring than putting a big weight on it and letting it drop. One of the most important features of the mooring was how far off the bottom the sound source would be positioned. Since the RAFOS floats for this study were ballasted to drift 600 meters below the surface, the final resting position of the beacon had to be at about 600 meters as well. Too shallow or too deep, and the sound emitted from the source might become trapped above or below the SOFAR channel and not be heard by the floats. This phenomenon was so well understood that submarines have been known to hide within water layers to evade enemy SONAR.

Many months ago Amy had calculated the depth of the water at each deployment site and had sections of mooring cable premeasured for each sound source. Most moorings were placed in water so deep that multiple spools of wire were required to allow the sound source to float within the relatively shallow SOFAR channel. Every spool was labeled with the mooring letter to which it belonged, and Amy's mooring technician directed the ship's boatswain and a deckhand in the correct loading order of one of the four-foot-diameter reels onto the pay-out winch.

The mooring was actually put over the side upside down; that is, the part that floated nearer the surface was cast off first. This part was a series of 10 large glass balls. These balls had a wall thickness of about half an inch, and this, combined with the inherent strength of a sphere shape under pressure, allowed them to provide buoyancy while surviving the crushing depth. These basketball-sized glass spheres were called "hard hats," because the yellow plastic protective enclosures that surrounded them resembled construction workers' hats. The hard hats were linked together with chain and provided flotation to keep the sound source upright off the sea floor.

It was still dark. Floodlights from various places above my head provided an almost shadowless illumination that reminded me of a night game at a ballpark. The deck moved up and down, tossed about by unseen waves. From where I stood I could see the string of hard hats being towed in *Knorr*'s wake. This ensured that the mooring cable would not become fouled on itself or around the ship's propeller. As more cable streamed off the fantail, the strain on the line became greater and the chance of an accident occurring on deck increased.

Wearing a real hard hat and a floatation vest, I carefully moved across the deck, sensitive to the bight of line that was holding the cable load. The mooring diagram that Ann was holding indicated that this was the point at which we needed to attach the sound beacon. Ann and I moved the unit into position. We were working a few feet from the edge and as *Knorr*'s stern rose and fell in the building seas, waves periodically broke over the transom and sent warm Indian Ocean water swirling around our ankles.

The boatswain stopped to read my inscription on the side of the housing. "That ought to raise some hackles if it's ever pulled up in a fishing net," he said.

I thought about this for a moment, took out my marker and scratched out what I had written on the yellow cylinder. Slim chance that it would ever be found, but I didn't want to be the start of some international incident.

We attached the sound source, working quickly to tighten the shackle connections. The boatswain looked around carefully. Only when everything met his standards of safety and workmanship did he raise his arm, and using hand signals, tell the winch operator to begin letting cable out. The three of us then eased the device over the fantail.

I turned around to see Amy standing in the shade of an overhanging deck, chewing the cuticle of her thumbnail. She was too far away to see the details of the mooring assembly, but she had the sequence committed to memory and was listening for any indication of trouble. Unlike a lot of chief scientists, Amy let her technicians do their jobs without double-checking every connection and measurement herself. Some of this control she was forced to relinquish because of her vision, but I think it was in large part a matter of her management style.

The next step was to attach 1000 meters of cable. A wooden spool labeled "Red Sox – mooring E" began to slowly unwind as the yellow hard hats drifted further away until they were barely visible, bobbing in the waves. We paid out more cable, and after the spool was finished, another cable was shackled to the first, all while the ship crept closer to the drop zone. Timing was everything and the objective was to have the last of the cable out just as the ship glided over the mooring area.

The drum stopped turning, and *Knorr's* boatswain and the mooring technician attached an old railroad-car wheel to the end of the cable. This low-tech but very heavy object would be the thing that kept the sound source secured to the bottom for the next few years. I heard the change in pitch as the ship's engines idled down, and the third mate's disembodied voice called out from the ship's PA system.

"Half mile to position E, half mile."

The boatswain looked skyward, contemplating something, and then turned to me.

"Call the bridge, tell them to slow down and take their time. We need a few more minutes."

I opened the stern phone box, called the mate, and relayed the information.

"Whatever you want, I've got no other plans for the morning," the mate answered me.

When everything was ready the deck crane lifted the wheel over the stern. In less than a minute the phone box rang. "We're here and speed is zero," the mate said over the intercom.

The boatswain held the release lever in one hand and stared down at the water for a few seconds as if he knew something the mate didn't about our position. "Stand clear," he said, and with a tug on the release lever, he dropped the wheel into the water. The resulting splash looked small after all the effort leading up to it.

Lost in a deep purple predawn twilight, the yellow balls were impossible to see from the ship, but in a few moments they would be pulled down to the deep. All I could do now was take a deep breath and look east toward the rising sun.

I walked over to where Amy was standing. "I guess in 12 months we'll know if it is really working."

"I'd like to be able to put a hydrophone over the side and wait for the next scheduled tone and hear it for myself, but we can't afford the time," she replied, no more satisfied than I was with the uncertainty of this whole process. There was always the chance that the increase in pressure during decent would cause a seal to fail, or the cold temperature at depth would sap the strength from the batteries, turning the whole assembly into a very expensive piece of underwater modern art.

With E hopefully moored safely to the bottom of the Gulf, the last of the scheduled work was completed. The mate wasted no time throttling up the ship's two 1500-horsepower engines to a cruising speed of 12 knots. A sharp turn rolled the ship to starboard as *Knorr*'s bow swung toward the Seychelles Islands, 1000 nautical miles southeast beyond the sunburned red horizon.

18

THE FOUR-DAY TRANSIT TO PORT meant an end to the watchstanding requirements of the science party. Amy confirmed this during breakfast.

"You can head to bed if you want. Your only responsibility between now and when we make port is to get everything packed up."

For me these words were a presidential pardon from my midnight-to-noon prison sentence. More than beer on demand, more than a nice big bed with thick pillows, more than anything else I had wanted to get my body back on its own circadian rhythm. Late to bed, late to rise, just as God intended.

"I'm not tired," I said.

I couldn't believe the words were coming out of my own mouth. Maybe I was finally getting comfortable missing supper; or maybe, being no longer required to force sleep at one in the afternoon, my rebellious nature sensed the loss of challenge. In any event I was definitely NOT tired, so after finishing my eggs I shuffled off to the lab and assessed the challenge of rounding up all our gear and equipment, which by now had diffused its way into every crevice of *Knorr*'s interior like a spreading vapor.

"Your job must be finished too," I said to Jim, who was sitting in front of one of the lab's computers checking his email.

"Yeah, time to relax. Not much likely to happen steaming along in the middle of the Indian Ocean," he said, not taking his eyes off the screen.

It would have been interesting to see a visible difference in Jim when he was "relaxed," but I didn't detect much change.

Now began the task of repacking gear and electronic equipment. With most of our expendable material dispersed around the four corners of the Gulf of Aden, there was extra room in a few of the shipping containers. In place of anchor chain, shackles, and other miscellaneous jettisoned RAFOS float paraphernalia, boxes were stuffed with sea boots, books, and bags of dirty clothes. Every time I was ready to close the lid on another box, someone came into the lab with a trash bag full of stuff.

"Do you have any extra space? I'm thinking of doing some hiking in the Seychelles and I don't want to lug my extra junk around," Jeff, a young graduate student from Woods Hole, asked me, his eyes full of hope.

It seems everyone had the same idea: travel light while enjoying a few vacation days on the islands. Why lug around a bunch of dirty underwear? These boxes would take more than a month to get home, with a large part of that time spent sitting in the scalding equatorial sun. The thought of opening them after the contents had fermented like ripe cheese was not a happy one, but as we approached the end of the cruise I was in an agreeable mood. If it fit in the box, in it went.

Two hours later, and after a long lesson on the compressibility of clothes, a call came over the ship's PA system. It was the voice of the third mate. "*Knorr* will be crossing zero degrees latitude in approximately 30 minutes. Anyone interested in driving some golf balls across the equator should muster on the fantail."

I never felt that hitting a golf ball from the southern hemisphere into the northern hemisphere was an experience lacking on my resume, but it was an experience not afforded to many other people in the world. So I dropped the trash bag in my hand and called it quits for the day.

I didn't walk straight to the fantail, but took a detour to my cabin to retrieve my Shellback card. The images and smells of our recent equator crossing indoctrination were still fresh in my mind. I did not want to take the chance that this was some lingering cruel joke.

Shellback card prominently displayed, and alert for anyone approaching me with a black olive or peanut butter, I emerged into the painfully bright light. A small group had already gathered. The boatswain had lain out a two-foot square of green Astroturf and removed *Knorr*'s stern railing. The captain emerged on deck with a full set of clubs in an expensive-looking bag. Each of the woods, attired in a knit bonnet with tufted pom-pom, looked overdressed for the 90-degree heat that rose up from the steel plate at my feet. We stood around as if it was the start of the British open, acting the part of a hushed gallery while waiting for the captain to select his club. The third mate's voice resonated around the ship again. "We have just crossed the line."

Knorr slowed down to headway speed. When the captain was satisfied that the ship was traveling at a speed conducive to a 3-wood, he stepped up to the green square of neatly trimmed plastic grass and placed his ball on a makeshift tee. The ship's wake, two frothy white lines, defined a fairway stretching to the horizon. Not a sand trap in sight.

The captain pulled back his club in slow motion and released it on a blinding arc to the ball. Crack! All eyes followed the ball's flight until it splashed a respectable length from the stern, right in the middle of the liquid fairway. With no objects behind us to help us gauge the distance, we could only guess how far the ball had traveled. All around me heads nodded in approval as the captain handed the club to the boatswain.

A few other members of the crew took their turn. Some slices and the occasional hook, but nothing rivaled the captain's drive. The chief mate pointed the shaft of a club in my direction. "Here, give it a whack," he said. With a growing audience I self-consciously stepped to the mat.

I golf once a year, if only to remind myself how bad I am at it. I bent over and teed up my ball before realizing we were probably violating a number of international treaties regarding the dumping of plastics into the sea. And then there was my golf phobia; water hazards. I stood up, looked over a vast expanse of blue, and mumbled to myself about confronting fears.

After a few very good-looking practice swings to loosen up, I heard a voice from somewhere behind me. "Are you going hit it or just threaten it all day?" With my quota of good swings now used up, I took a half-step closer to the ball and set chaos in motion; hands, arms, hips, and legs went in different directions. The club head struck the ball well off center and the resulting trajectory defied the laws of physics. The tee was only eight inches from the deep blue sea. A puff of wind would have sent the ball harmlessly over the stern. Instead, what could be heard (the ball was flying too fast to actually see) was the sound of a dimpled plastic sphere ricocheting off a metal bulkhead inches from the captain's ear. This performance elicited laughter from the crew, profuse sweating from the captain's forehead, and the chief mate's probable thought that he'd come within a hair's breadth of promotion. Avoiding the captain's eyes, I handed the club back to the mate, wondering if walking the plank was still the traditional punishment for killing a captain with his own golf ball.

19

THE NEXT DAY AFTER LUNCH, a smudge appeared on the horizon. Above the smudge were the only clouds visible for as far as I could see, the tell-tale sign of an island. A warm, moist land mass surrounded by cool water causes the air over the land to rise. As the air rises it condenses into water droplets forming clouds that fly like a flag over far-off islands, beckoning ships to stop and visit. Sailors since the beginning of ocean voyaging have learned to recognize this sign and use it to find tiny dots of land in the vastness of the ocean.

On *Knorr*'s bridge the RADAR showed these islands from a distance of about 50 miles. The GPS navigator could point us to this specific harbor from halfway around the globe, but it was the sight of those clouds that confirmed to me that our cruise was almost over.

As we got closer the low smudge rose up to become a high steep mountain where lush vegetation sloped to a cloud-ringed summit. The green of the foliage was so vivid it was as if someone had turned the color adjustment on the TV to max. Before me lay a postcard paradise. The colors of the island, combined with the cloying earthy scent of land, contrasted starkly with the endless blue of ocean and the smell of diesel fumes of the last 40 days.

Above the bridge, the yellow quarantine signal flag, required of all ships entering foreign ports, cracked in the fragrant breeze.

Below the yellow square, the flag of the Seychelles flew in recognition of our destination. A VHF radio call to customs officials ashore announced that we were requesting clearance; *Knorr* was told to drop anchor in the outer harbor until clearance was complete. My romanticized visions of sailing into a distant bay had been influenced by old movies in which tramp steamers charged into port, barely stopping before letting loose 90-foot shots of chain through the hawse pipe. There was a clattering racket as the anchor dropped to the sandy bottom. In reality "dropped" was somewhat of a misnomer, as the massive anchor hung for quite some time with water lapping at its flukes while the boatswain pounded rusted chain links with a sledge hammer. After half an hour the anchor moved toward the bottom as chain crept down the hole at the speed of cold molasses. Our patience was rewarded in the end when we were able to see the anchor on the bottom, clearly visible through 60 feet of water: an advertisement for some potentially great diving.

With nothing to do but wait, Amy and I thumbed through a Seychelles guidebook someone had brought aboard. Everything was packed and the dock was in sight, and Amy finally let herself relax, content to leave the matter of clearing-in to the ship's officers. As we sat in the library, I read some selected facts about the country. Located truly in the middle of Indian-Ocean-Nowhere, the 115 individual islands that comprise the Seychelles were spread over an area about the size of California. We had arrived on the main island of Mahe. To get to any other country from here we would have had to travel at least 2800 kilometers to India or 1600 kilometers to Africa. This isolation gave the islands a rather unique flora and fauna, with a lot of species found no place else on earth, and Amy and I were eager to see some of it.

It was late in the afternoon and I was beginning to wonder if we would ever be allowed ashore. One after another, small boats pulled up alongside *Knorr* to discharge their official Bureaucratii cargo. I leaned over the rail and started a conversation with one of the men attending the latest launch to tie alongside.

"Does it usually take this long to get cleared in?" I asked.

He looked up at me, and then to the flag halyard high above our heads. With a stern face he replied, "That is not the flag of the Seychelles. We have not flown those colors in over three years."

There was no law requiring the flying of the visited country's flag, but the courtesy was apparently appreciated in some countries more than others. Only after being boarded by representatives from the Port Office, Immigration, Customs, The Department of Agriculture, The ministry of Health, and the Harbor Pilot were we allowed proceed to our berth on the quay.

The process of tying up almost 300 feet of ship was painfully slow, more so after so much time at sea. I couldn't wait to walk on grass or dirt or even asphalt, anything but steel deck. I wasn't alone. As soon as the gangway was lowered to the dock a large group of us hurried down its steep incline. The solid ground felt odd under foot. Since entering Mahe harbor *Knorr*'s movement had been imperceptible, even nonexistent, but there was still a difference to be felt as Amy and I walked toward the port gate. Each step transmitted a sensation of substance through my soles, an intimate contact with the earth that I had missed aboard *Knorr.*

"So what do you want to do first?" I asked Amy.

"A cold beer. After that, I want to walk for more than 50 paces without having to climb a ladder," she said, snapping open her folding red-tipped cane—the cane that had not seen the light of day on the ship.

Later that evening Amy and I stood in front of "Victoria's most famous bar"—The Pirates' Arms. If by famous was meant loud, rowdy, and expensive, then the advertising was truthful. A long row of windows were open to the street, and even though it was a moonless night, bright, fluorescent light from the awning lit up the sidewalk as if it was day. Dozens of people, dressed mostly in tee shirts and shorts, walked by. Most stopped to look in, as if to test the mood of the place, and a few passed through the open set of double doors. Inside, *Knorr*'s entire crew and science party were gathered for the kind of drinking people can only do after being away from alcohol for a couple of months. The smells of pizza, spicy Indian dishes, and tobacco smoke wafted together, while a large flashing sign on the back wall announced the current jackpot of the casino in the back room. One of the engine department's oilers, a man with a belly the size of Vermont, was standing on top of a table. I had not heard him make an audible noise since I'd stepped aboard *Knorr,* but there he was, three feet off the floor howling in a credible impression of a coyote. The odd thing was that he was attracting no attention.

When the waitress began carrying back more liquids than solids, Amy and Steve hastily signaled for the check and moved off to a relatively quiet corner to look over the long food and bar tab. It was customary for the expedition leaders of a research cruise to pay for the first dinner ashore, and judging by the long scroll of paper Steve had in his hand, it was not going to be cheap. I saw Steve straining to be heard over the din of music and voices as he read the list to Amy. I was impressed by how easily the rest of the crew did their best to accommodate Amy's vision loss, and at times like this I didn't feel the need to rush over and help.

We staggered back to the ship for our last sleep afloat. Some of the off-going crew and members of the science party would be flying home the next morning, but most of us had plans to spend a few days touring the island and relaxing on a beach. Preferably a private beach. I think the rest of the crew shared my sentiments. They were a great group of people, but after 40 days, everyone longed for some alone time.

The next morning dawned loudly and the sun burned through the porthole with the intensity of a laser. I rolled out of my bunk dressed in yesterday's clothes. The source of the noisy dawn was between my ears; a steady pounding timed with my heartbeat. I had never tolerated alcohol well, and abstaining from it for so long made the previous night's moderate drinking all the more toxic. Shielding my eyes from the light saber poking through the porthole, I scanned the cabin to find Amy already gone.

In the time it took me to walk down the passageway to the galley, the ship came to life, though at a much slower pace than would be considered acceptable at sea. Some cabin doors were open; occupants were packing their gear or cleaning up in preparation for a quick departure. Later, *Knorr* would provision, take on fuel, and swap containers of scientific equipment in preparation for the next leg, scheduled to begin in a few days. The smell of bacon and coffee acted like an invisible rope pulling me to breakfast. By the time I reached the mess deck the pounding in my ears had subsided to an uncomfortable dull heaviness. After a quick meal, Amy and I made one last sweep of the lab to make sure all the shipping crates were ready to be off-loaded.

The logistics of coordinating supplies, people, and machinery through foreign ports halfway around the world required an immense amount of preparation on the part of the scientists and the WHOI Port Office. It also required the services of a port agent who knew the nuances of the country's bureaucracy. Amy and I were to meet our agent on the quay. The ship's crane was already lowering plastic crates to the ground as Francois, a short trim man wearing slacks and a tropical print shirt, pulled up in an open jeep. Behind him, three much younger men in similar attire sat quietly, crammed in the back seat.

"Bon morning," Francois greeted us, in an amalgam of French and English. This was probably the result of alternating French and British rule prior to Seychelles independence in 1976. We made our introductions and handed over a manifest of the contents of each box.

"Don't worry about a thing," Francois said in a too-relaxed manner that suggested we should worry about everything. "It will all be taken care of. Now you go and enjoy your vacation on our beautiful islands."

He was right, of course. There was nothing left for us to do. It was here we gave up control and set our watches to Island time, which meant take them off and leave them on the nightstand. With attitudes properly adjusted, Amy and I made our way to downtown Victoria. At an ATM we discovered that though the Seychellois did not keep time it did keep count; relaxation came with a price. The official exchange rate was about six rupees to the dollar. Upon leaving the bank, we bumped into the third mate, who told us he'd just changed money at a Chinese restaurant up the street at 14 rupees to the dollar. Sensing counterfeiting, money laundering, or racketeering, we opted to stick with the ATM, but our concern was costly. It seemed that everyone changed money on the black market and if we didn't follow suit our stay in paradise would be twice as expensive.

Money in hand, we loaded our bags into a white Suzuki Samurai that we'd rented. Despite the warm sunny weather, with all our belongings in the back we thought it prudent to keep the black canvas top up and the car lockable. Soon we were negotiating the small streets of Victoria in search of the main road out of town.

After dozens of wrong turns and dead-end streets, I was feeling intimidated by the deceptively small size of Victoria. Nor was the Samurai helping my confidence behind the wheel. This car should have been named the Suzuki Kamikaze, since it was forever trying to commit suicide by shedding mechanical parts when I so much as closed the glove box. To the car's credit, the tailpipe emissions looked incredibly clean when compared to the belching smog of most other cars on the road. It wasn't until later that I learned this was due to the tailpipe not being connected to the engine. All of the exhaust gases were being filtered first through the passenger compartment. This was probably why the Seychelles' tourism board described driving around Mahe as ". . . an intoxicating experience that sets your mind swimming. . . ." Given a choice between asphyxiation and loss of our stuff, we decided to drop the top and enjoy the scenery with clear heads.

One legacy from British rule was driving on the left side of the road. Amy did her part by reciting the stay-alive mantra at every intersection or rotary—"Look right, stay left"—as we motored along to the south end of the island. Further from Victoria, large buildings gave way to small houses and then to fishing shacks, and finally we were enveloped in thick, green foliage.

Mahe had two main north/south roads that were connected by a half-dozen east/west ribbons of pavement that crossed over a high mountainous spine in the middle. The roads themselves were in good shape and could be considered acceptable as single-lane, secondary roads in flat Nebraska corn country; but add two lanes of traffic, hairpin turns, no guard rails, 500-foot cliffs, pedestrians, and the odd goat, and the drive was more like something only to be done on a dare. The south end was much less populated than the north, which I suspected was a Darwinian phenomena; few people survive the drive south. Each time I flogged the Suzuki up a steep mountain pass the rearview mirror was filled with the image of a bus grill. Old school busses billowed black smoke for their entire ascent, until they crested the top, at which point the smell of burning brake pads followed their manic trip to the bottom.

We arrived at our hotel after having successfully navigated the roads without killing anything of consequence (I don't count the chicken, she was probably going to be eaten anyway), and I finally allowed myself time to take in the view. Our room, built into the steep hillside, looked out over the tops of flowering hibiscus trees to a crescent-shaped beach. The beach was deserted and the turquoise water contrasted sharply with white sand and gray volcanic rock along the shore. It felt good to sit on the shaded verandah and appreciate the ocean from a distance for a change.

"If I'd known that this is what being married to an oceanographer was like I would have done it a long time ago," I said as I wiped condensation off my beer with an index finger.

"A long time ago you wouldn't have had many women oceanographers to choose from," Amy replied. "Besides, most spouses of oceanographers stay home and take care of the kids."

A few months ago we'd talked about the idea of talking about kids on this cruise. For all our time together on those eight weeks at sea, the topic had never came up. I don't know if that was because of the constant distractions of the work at hand, or because we were avoiding the difficult details of the subject. Our schedules were packed with both professional and personal commitments. We were just coming into that stage in our lives where we had the freedom to do whatever we wanted; travel, hobbies, careers. Were we ready to give that up for school plays, soccer games, and snotty noses? Amy obviously didn't have a typical job. Her being away from home for long periods of time would be difficult for all of us, but most of all for an infant who would not understand why Mommy suddenly disappeared. Then there was the biggest question of all. The big gorilla in the room we tried so hard to ignore. What would it be like to try and accomplish all of this if her vision got worse? There were scary things to consider. How would Amy know if the baby was putting something dangerous in her mouth? What would happen if she toddled off out of Amy's range of sight? How do you sit and read a bedtime book with your child if you can't see the words or point to the pictures?

We had discussed some of these questions before, but never really sought the answers. All of our conversations about starting a family had been very analytical; the pros on one side of the page, the cons on the other. In every instance we could never come to a conclusion. There was always some fundamental emotional component missing to push us to a decision. As we sat there under wind-swept palms looking over infinite shades of blue water, we were still in no position to mull over dirty diapers and daycare. Any serious consideration of children would have to wait until we got home. The only thing we could decide on was that we were hungry.

After shelter, the only other requisite elements to vacation survival are food and drink. We detoured on our way along the beach to the nearby Anchor Café. The Anchor Café was a small shack with a couple of picnic tables out front. Its namesake anchor, a monstrous hunk of metal that looked like it was retired from the aircraft carrier *Nimitz*, lay half buried on the front lawn.

Amy and I ordered burgers that in their previous life probably ran the Kentucky Derby once or twice. We were the only patrons and we ate slowly in the shade of the anchor while talking about the events of the past few weeks. I was into my last bite of Man-O-War when a man in bare feet and a faded Rolling Stones '79 Tour tee shirt stepped out of the café's side door. He took the bench seat opposite us on the picnic table.

"Name's Tom." He extended his hand eagerly to both of us. "So you two are off that research ship."

My face must have looked like one big question mark because he continued immediately without waiting for an answer. "I heard you talking, we don't see too many Americans here." It was then that I realized he was also an American. It turned out he was the owner of the Café.

"How did you happen to land in the Seychelles?" I asked.

"Came here 25 years ago to work on a U.S. satellite tracking station. Fell in love with the island and out of love with my job. The only thing I track now is rupees going into the cash register."

"I know how you feel. I was in a similar line of work," I said, dipping a soggy fry into an odd-tasting sauce.

"It's good to hear a Yankee voice. We get so few Americans here I almost forget what I sound like," he continued, seeming anxious to make conversation.

"There's a shipload of us running around here somewhere," Amy said.

"Yeah, I know. Your captain stopped by earlier today. He was looking for a fruit bat to deliver to the ship's cook."

"Why would he want a bat?" I figured this was some sort of joke.

"Most people on the island eat bat," Tom replied.

I looked down at the crumbs on my plate. "You don't say?" It was the only phrase I could think of quickly that didn't sound judgmental. He seemed to interpret my question as continued interest.

"My children keep a fruit bat as a pet. It's in the backyard. Would you like to see it?"

Without waiting for an answer Tom got to his feet and motioned for us to follow him across the coarse grass lawn. Behind the restaurant stood a modest single-story house and a cage containing a furry brown bat the size of a seagull. The bat hung upside down, head tucked under a wing. It moved its wing slightly to get a better view of the people so rudely interrupting a good day's sleep. I could sympathize, given the erratic sleep schedule forced upon me for the last two months.

"They don't like sunlight," said Tom.

I wanted to say "anyone who has seen a vampire movie knows that, tell us something new," but instead I nodded and let him continue his lecture on the life of a Seychelles fruit bat.

"Not much fun as a pet. It doesn't look like it does much during the day," Amy said, leaning close to the enclosure for a better view.

"It's perfect if you're an insomniac. Heck, it's no different than a hamster," I said, wedging my finger between the wires to scratch the bat's head.

"Watch out for your finger. He likes fingers," came a small voice from behind us. A young girl, about nine years old, joined us. Her frizzy dark hair was matted with sand and saltwater. She stood barefoot, wearing only a bathing suit bottom, her deeply tanned skin also flecked with sand.

"This is my daughter Anika," said Tom. "It's her bat."

"What do you feed him?" Amy asked.

"He likes bananas, but mangoes are his favorite," she replied, sticking her finger in the cage as I had done, to scratch the bat's head.

Amy continued her dialog with Anika, drawing out answers to her questions and showing a real interest in the girl's responses. I could tell then and there that any child would be lucky to have Amy as a mother.

As we turned to leave, the bat slung a wing back over his head, much less enthusiastic about having company than Tom and his daughter were.

After our bat-burger experience at the Anchor Café Amy and

I talked again about setting aside some time to have a discussion about kids, and also vowed to be more selective about our future dining plans. A vow that would have to wait another meal, because later that evening we were scheduled to meet some shipmates for dinner at a place we hadn't chosen. All we had been given was the name of the restaurant and a vague notion of its location. Later that day we drove the unlit streets under a darkening sky, but the place proved elusive. The search was made more difficult by the fact that the Samurai was also a Cyclops. Only the beam of a singled headlamp illuminated the road.

I was standing on a dimly lit street corner contemplating the one piece of advice that should be provided to anyone planning to visit the Seychelles—never ask a local for directions. Though they lived on an island you could throw a rock across, they couldn't locate a urinal while standing in a men's room.

"Excuse me, can you tell us how to get to Le' Maison Charles?" I asked a small group of young men standing outside a convenience store.

"I think it's closed," said a man with a gold-capped front tooth.

"No, no, it's on the west side of the island. I know, my sister-in-law ate there once," said the man sitting next to him.

A third man walked out of the bushes behind the two, zipping up his fly. "Don't listen to them. It is up this road a few kilometers on the left," he said, picking up an unfinished beer. He seemed the least intoxicated of the three, so we assumed his advice to be the most reliable.

We soon passed the point of being fashionably late; and thus far, none of the directions we had been given had gotten us any closer to our dinner engagement. After another 30 minutes of aimless driving, we at last found the restaurant. The entrance was 100 feet from the convenience store.

After all those weeks in the confined spaces of *Knorr* the ability to get lost was somewhat liberating. The structured and scheduled routine served its purpose at sea but was in sharp contrast to the ultimate flexibility we enjoyed basking on the beach, or wandering aimlessly around town. After a few days of snorkeling, sunning, and eating our way through this Nirvana, Amy and I were ready to get home. I had put my business on hold for more than eight weeks and it was time to get back and find out if there was anything left of it.

"So are you willing to do this again? The second half of this study is in six months," Amy said. "There's still time to change your mind."

With two cruises under my belt and the opportunity to come back to Africa again in the not-too-distant future, I could think of nothing else I would rather do. The business would survive, or not. I wasn't driven to make it bigger, at least not at the expense of some great life experiences. "If you still need me to go I wouldn't miss it for anything," I said, without exaggeration.

Amy fingered a bead necklace I'd bought her in Mombassa. "I'll be running the next one on my own," she said. "I'm going to need a lot more help. Not to mention it's going to be on a ship I've never seen. To be honest, I'm a little nervous about being able to pull it off.

I could tell she was getting into one of her moods, and she knew it too. Amy was all too self-aware of what she called her "Imposter Syndrome." It was something she'd read about and identified with. The idea is that some people, mostly women, feel that they didn't really earn their success, that it must have all been luck, a luck which could run out at any time. I knew Amy felt this way. It didn't matter that she was a tenured scientist at one of the world's most prestigious ocean research institutions; she still carried a disclaimer in her head, one that repeated the investment-fund mantra "past performance is no guarantee of future returns." For Amy, the constant battle with deteriorating vision only reinforced this perception of precarious success. Coming off the high of just completing a very successful and challenging project only made the potential fall that much more ominous.

"It will be fine," I said, fully believing that I was right. "You got all the bugs worked out on this trip, the next one will be a piece of cake." "Piece of cake" would turn out to be quite a misnomer for that still-distant cruise; but I had no way of knowing that.

With great expectations of another adventure in the near future, we threw our bags in the back of the Suzuki, which to my surprise had retained enough of its components to get us to the airport for the long series of flights home.

20

SIX MONTHS HAD PASSED since our return from the Seychelles and *Knorr.* Amy had spent the time preparing for the second half of her Red Sea study. We were to repeat the same cruise track, but in the summer months. The question to answer was this: do seasonal variations in sea temperature and wind affect the circulation of water flowing from the Red Sea into the Indian Ocean? In the winter, monsoon winds of the region whip down from the Arabian Peninsula to the north. In summer, the monsoon reverses and the wind originates over the African continent. Amy's hope was that we would see the effects of the summer winds, and not spend the time drifting on a hot, glass-like sea.

It was one day before we were scheduled to depart for Djibouti, a tiny country on the coast of northern Africa. Air France had just called to say our flight had been canceled. The airline representative spoke in an apologetic French tone usually reserved for times when the Brie was served too cold. He would, he assured us, put Amy and me on the next available flight to Paris later that evening. The only problem was that the later flight allowed only a 30-minute connection for our leg to Djibouti. "But of course you will make your connection, monsieur. This is Air France, we take our schedule seriously."

I knew there was no better guarantee that our departing flight would be delayed. Sure enough, an hour after our scheduled departure time we pushed back from the gate at Logan Airport in Boston. Once in the air I performed a traveler's version of masochism by asking the flight attendant if it was likely that we would make our connection in Paris. "But of course," came her reply, using the quaint phrase in a familiar tone that meant "What a stupid question you fool. You must have an I.Q. lower than your seat number."

Halfway across the Atlantic I found myself pondering String Theory. String Theory is an unexplainable astrophysical concept that seeks to explain all of those other unexplainable things that happen in the universe. Somewhere lost in the convoluted equation of this theory lies the reason why connecting flights depart on time only when the originating flight is delayed. We arrived in Paris to hear our names announced over the plane's PA system. We had 15 minutes to make our connection.

The instructions that followed were lost in a combination of thickly accented French/English and a poor quality audio system found only on $500 million aircraft and fast-food drive-throughs. We were ushered past other de-planing passengers through the plane's galley door and into a waiting van.

As we boarded the van I asked the driver if our luggage would also make it in time. "But of course (you moron!)," came his reply. I slumped into the seat with a hesitant feeling of relief and stared blankly out the window. My body, sleep deprived from the overnight trans-Atlantic flight, was tossed around like a limp doll as the van careened across the tarmac.

Ten minutes later a woman with a voice as sharp as her thin, pointed nose roused us into a higher state of consciousness. "Quickly, quickly. Take your things and follow me."

We ran through the terminal building, bags in tow, trying to keep the pointy woman in sight. After what seemed like the end of a marathon we found her standing at a gate next to a second agent holding out a hand for our boarding passes. I handed the agent our tickets. "Will our luggage make this flight?" I asked once again.

Her condescending glare saved her the breath of saying, "but of course!" Once through the jetway, we boarded, not a plane, but another van for a mad dash across what was beginning to look like very familiar scenery. The van sped past the plane we'd just disembarked from and pulled up to an adjacent aircraft waiting to take us to Djibouti.

Buckling my seat belt, I looked out the window at the plane we'd arrived on, and thanked God that even though we had just circumnavigated Charles de Gaulle airport our luggage had apparently only traveled 200 feet. I felt confident we'd have clean underwear in Africa.

The flight to Djibouti made a stop in Jeddah, Saudi Arabia. Saudi law dictates that all alcohol on the plane be locked up once the plane enters Saudi airspace. Even empty wine bottles have to be removed from the seat backs prior to landing. Two Saudi alcohol inspectors with tobacco-stained fingers made their way up the aisle, checking between cushions and under seats for any drop of the demon rum. It seems you can chain-smoke unfiltered Marlboros in Riyadh all day, but swallow a little alcohol-tinged mouthwash and you may lose your head in the village square.

After 24 hours of travel we arrived in Djibouti. I lay on the hotel bed wearing a new tee shirt. It was the only article of clean clothing I had, and it said *compliments of Air France* on the tag. It should have said "*condolences of Air France*," as all of our luggage was still in Paris.

"But of course!" I said to the ceiling.

"What?" Amy asked from the bathroom.

"Nothing." I was too defeated to rehash our ordeal.

The next flight wasn't until the day after tomorrow; and with any luck, I thought, our luggage will be on it.

Our hotel could have been the Holiday Inn, Newark, with floral print paper on the walls and a distinct odor of mildew, the result of marginal air-conditioning. Splintered rattan furniture sat on a carpet the true color of which couldn't be discerned through the layers of stains.

I picked up the phone to call Air France on the odd chance

that our luggage had been misplaced in Djibouti. The phone had no dial tone. Tracing back the cord I found four splayed ends of bare copper wire. Looking at four similarly bare wires protruding from the wall, I figured there were at most 24 possible connection combinations. While I worked on our communication problem, Amy tried to figure out why water came out of the showerhead whenever she opened the sink faucet.

The next morning we took a taxi to the Air France office. Djibouti made Mombassa look like a modern metropolis. Goats meandered the city streets looking for scraps. A layer of dust covered everything and everyone. We hoped to conduct our business and get back to the hotel by noon, not so much for lunch, but to avoid being on the streets later in the day. According to our guidebook, every afternoon a flight arrived from Ethiopia loaded with quat, a leafy narcotic that the locals chewed. By three o'clock most of the cab drivers were supposedly high on the stuff. The taxi we took from the hotel had doors held closed with duct tape, and I didn't want to test the quality of that latch by putting my life in the hands of a stoned driver.

It was still early though, and armed with a fist full of local currency, "compliments" of Air France, we visited the open market in a desperate search for clothes to replace our still-unaccounted-for wardrobe.

"For you my friend, best price," said one shopkeeper with an arm full of colorful scarves.

"My friend, my friend, excellent watches . . . Rolex, best price," said another from across the path.

It seemed everyone in Djibouti was my friend, but unlike my friends at home these friends wanted my money. It was soon apparent that we were no match for these vendors, for whom the art of negotiation was a birthright. What made it fun was that we were bargaining with the airline's money, so we got to play the game a little differently. A typical exchange went like this: "How much for that tee shirt? . . . $10, are you crazy? . . . That's much too low. Here's $12." The fabric of Middle-Eastern trading society may have been forever torn by our exploits in the bazaar. Some vendors threw up their hands in disgust at our inept trading practices.

Clothed in the latest designer knock-offs, we met the ship's agent, who was to deliver us to the ship for a few hours of preparation work. We wouldn't be able to move onto the ship for another day. Arriving at the port office, we were asked to surrender our passports to an unsmiling immigration officer before boarding the ship. Being separated from my passport was like having a tooth pulled without Novocain; I didn't do it without a lot of pain and anguish. But our agent assured me everything would be fine.

From the port office we drove down a crumbling stone quay. The R.V. *Raven* lay alongside with an aluminum gangway leading up to the main deck. *Raven* was owned by the National Science Foundation and like *Knorr* was part of the UNOLS fleet of research vessels. The daily operation of the ship was the responsibility of the Northeast Ocean Institute in New York, and much of her crew hailed from there. The 230-foot *Raven* was listed as Baltic Ice Class IA, which, as I stood 12 degrees north of the equator in 95-degree heat, seemed a bit excessive.

Fifty feet smaller than *Knorr*, *Raven* was still an imposing ship when viewed from the dock. Her white superstructure reflected the brilliance of the African sun and I had to squint to see the blue crown painted on the smokestack, a strange emblem for an American vessel. Amy and I reported aboard, then spent the next hour walking the passageways so that she could get the lay of the ship.

"I think I have a pretty good idea of where things are. What I don't know should come pretty quickly," Amy finally said to me.

I was a little surprised; one of Amy's biggest concerns in the last few years of diminishing vision had been wondering how well she would get around on the research ships. If she had any concerns about this ship, she didn't tell me.

By then the remainder of the science party was aboard. Some familiar faces were from the *Knorr* cruise, but there were a number of new ones in the form of graduate students. Amy organized the group and we spent the remainder of the day doing an inventory of equipment as boxes were unpacked. As busy as we were, there was little sign of the ship's crew. Those that passed through the lab space did so quickly and quietly as if not wanting to draw attention to themselves.

With work for the day complete we piled into the agent's van for a ride back to the hotel. After proceeding a hundred meters up the quay the van driver pulled up to a group of waiting cabs. None of the drivers were in their cars, which sported a random jumble of dented fenders and peeling paint. The men vied for the small patch of shade available under a lone tree near the port gate. As our van slowed to a stop the cab drivers rose in unison and swarmed around the front of our van. A dozen thin figures in dusty white robes began a verbal exchange with our driver that had every man screaming at the top of his lungs in a combination of French and Arabic. The six of us from the ship looked silently at one another. I thought maybe it was the quat talking, and expected fists to be flying through the window at any moment. I didn't understand the words, but could guess at the intent, and I'm pretty sure that given the gesticulating and volume there was a lot of "Your mother smells like a camel" and "Your sister looks like a goat."

Just as quickly as the cacophony started, it was over, with apparently nothing resolved. "They say I am stealing their business," our driver said over his shoulder as he put the van in gear and rolled about 50 feet to the port immigration building. An immigration officer wearing no identification except a white shirt with simple red epaulets emerged from the office door. As he approached the van our driver proceeded to yell at him for no obvious reason and with much the same venom that he'd used on the cab owners. After 20 minutes of hand waving, palm slapping, and screaming, the driver explained to us in halting French that our passports were now in the agent's downtown office. One member of our party had the good sense to refuse to leave the port until the agent delivered our passports to us. This led more ranting on the part of both the driver and the officer, the result of which confirmed that our passports were not in town at all, but back on the ship.

We drove back down the quay, retrieved our documents, and hopped in the van for what I thought was, at long last, a ride back to the hotel. What happened next was like a scene from an old Marx Brothers movie; the van again pulled up to the group of cab drivers, 10 minutes of requisite yelling and swearing ensued, and then the van rolled the same remaining 50 feet to the immigration office where, I kid you not, we were asked to surrender our passports. At that point I felt I could yell and scream as well as our driver, but decided against it so as not to add a Djibouti jail to my list of travel experiences.

Inside the derelict immigration building the entire science party handed over their only tangible link to the U.S. of A., and with a flourish of pounding, ink stamping, and yelling, our passports were returned to us along with a paper Shore Pass that we discovered, upon a close look, said we had just arrived from Athens, Greece. I now had two entry stamps into Djibouti with no sign that I'd ever left the county.

We arrived back at the hotel to find that our luggage had completed its little sojourn in Paris and decided to join us for the rest of the trip. The two suitcases sat huddled in the corner of the room looking like scolded dogs. I immediately opened mine up and began the process of getting reacquainted with my toothbrush.

Sometime during that night, I contracted what I later nicknamed my "Djibouti souvenir parasite": bad sore throat, high fever, and a thorough intestinal cleansing. The next morning, defying my nausea, I joined the rest of the group on one last van ride to the ship. Upon reaching the port gate the guard and our driver entered into the all-too-familiar shouting match, with the guard presumably screaming, "Allah himself could not pass through these gates today." After more than a few minutes of this I had reached my nausea limit and, at the risk of throwing up on the guard's shoes, thrust my Shore Pass through the open window. With a "why-didn't-you-say-so-before" look, the guard acknowledged every person in the van, each of whom was now waving an official-looking, if not accurate, shore pass in the air.

Finally back on the ship, I staggered to our cabin. Amy's position as chief scientist on this cruise allowed her to use the chief scientist cabin—a spacious suite on the upper deck whose size was at least four times larger than that of any cabin I'd inhabited before. Since sleeping with the chief scientist by default allowed me to partake of this relative luxury, I hauled our gear into the space we would call home for the next five weeks.

The décor was utilitarian: steel and Formica, all secured strongly to the brown-carpeted deck. Our cabin had the benefit not only of portholes but of a large viewing window looking out over the bow. In addition, the spacious cabin was equipped with a private head, refrigerator, a large sofa, and a television. As an indicator that these amenities were only to facilitate work, the sleeping arrangements were no different than any others aboard ship: two stacked bunks tucked into one corner of the room.

I walked into the hall to retrieve the last of my bags and noticed that the captain's cabin was directly across from ours. The door was open, so I stuck my still-feverish head in for a look. His cabin mirrored our own with the exception of a missing top bunk and a heavy, steel safe on the floor under the desk. It seemed the captain was not expected to share a room under any circumstances. The cabin also had the feel of being lived-in for some time, with small personal items like pictures and car models, the result of its occupant trying to replicate a real home.

With the heat boiling my brain and my stomach on the verge of another purge, I dragged my bag back into our cabin and crawled into my bunk, selecting the lower one without asking Amy's preference. She had left in search of the captain to report on my dismal condition.

I awoke with a start. My watch read 10 o'clock. That meant I had been asleep for almost four hours—and had missed dinner. I held on to the upper bunk to steady myself as I stood, and carefully picked my way across the stained grey carpet. I opened the nearest porthole and flooded the room with sunlight. It wasn't

10 p.m., it was 10 a.m. I had slept for more than 16 hours. The ship rolled gently under my feet. The sense of motion didn't come from any bacteria playing with my inner ear; *Raven* was under way, gliding up and over a gentle swell. The view from the porthole was nothing but ocean to the horizon. I had slept through our embarkation from Djibuti. Before this could fully register, Amy walked through the door.

"Feeling any better?" she asked.

"A little. But I can't imagine eating anything. And the sight of the toilet evokes some pretty bad memories."

Amy held out a pitcher of yellow-colored liquid, the shade of which is found nowhere in nature. "The captain asked the steward to make you some Gatorade. He's concerned that in this heat you'll get dehydrated. Don't worry about getting down to the lab for work. We've got everything covered."

I knew she didn't mean it that way, but to me it sounded like I was not needed, or worse, that somebody in the group was not only having to do his own job, but mine as well. I poured a glass of the iridescent, synthetic replacement bodily fluid and swallowed two large mouthfuls. Any intentions of bounding down to the lab to remedy my job situation were quickly overruled by a wave of nausea. I retreated back into my bunk, my counterattack defeated.

An hour later the captain paid me a visit.

"What I'd like to do is record your vital signs and pass all the information off to MAS," he said in a drawl containing a hint of Louisiana or Mississippi. MAS—Medical Advisory Systems—was a private company that provided medical assistance to ships at sea via satellite FAX. With this service at his disposal a captain's only qualifications as proxy physician were knowledge of basic first aid and the ability to dial a phone.

He took my temperature and blood pressure, got me to say "ahhh," and asked a long list of questions from a standardized form.

"I suggest you hang out here and take it easy tomorrow. There's no rush to get down to work," he said, his tone easy and laid back.

Regardless of what the captain thought, I couldn't help but feel like a prisoner confined to house-arrest.

The next morning I lifted myself out of bed with the idea of at least being seen in the lab. Even before my spine became vertical my head began to spin and things got blurry. I managed a few tentative steps before flopping down on the couch. Taking the captain's advice, I tried to make myself comfortable and turned on the television. To my surprise it was receiving the current movie playing in the lounge. It had just started, so I settled in for what I thought would be a couple of hours of killing time.

Within 20 minutes the movie abruptly stopped. The clock on the wall indicated it was almost time for the change of watch. The person in the lounge was probably going on duty and had pulled the VHS tape from the player. Not feeling up to a trek to the lounge to select my own film, I switched channels. A screen appeared showing the ship's speed and position, and the air temperature, wind, and other navigational data. The next channel was a graphical representation of the ship's track on the course line and the time to the next station. A few more clicks on the remote and I was looking at the deck outside the main lab. It was an image without sound. Two people hovered over the rosette, preparing to launch it over the side as water silently moved past the hull. Concurrent with the image on the screen I heard the engines come to idle speed and felt the ship decelerate.

I watched the operations on deck for a few moments. Though there was no audio there was no question that the figures working on deck were laughing and bantering back and forth. I should have been out there with them, getting into the rhythm of the work. It was going to be awkward for me to disrupt that process when I got cleared for duty. I didn't want to think about it so I switched back to the VCR channel; nothing but a blue screen—no movie. I had come halfway around the world and was floating between Africa and Asia only to sit on a couch and channel surf. With Bruce Springsteen's "57 channels and nothing on" playing in my head, I drank another glass of amber elixir and decided to make a second attempt at getting to the lab.

Before I could make it to the door the captain knocked and walked in. He was well over six feet tall and stepped into the cabin with a stooping gait typical of tall men who worked in confined spaces all day, ducking to avoid hitting their heads on something.

"I just got this back from MAS." He handed me a faintly printed TELEX page. It read:

> Symptoms indicate probable bacterial infection. Start antibiotics as listed below. Our immediate concern is not the infection, but the low pulse rate. Advise ship seek closest port of call and transfer patient to facility capable of performing EKG.

I had always been proud of my low resting heart rate, the effect, I assumed, of staying in at least minimally good aerobic condition.

"This sounds like they're ready to fit me with a pacemaker," I said. "I don't think we have to worry about my pulse."

The captain's expression did not change. He had the face of a person who was bothered by nothing, including the prospect of detouring the ship to the nearest qualified hospital (qualified meaning one that wasn't run by warlords or despots), which in our case would have been almost 2000 miles away.

"Really, I'm okay," I insisted.

"As long as you're fine with it I have no problem. Here's your medicine." He handed me a small paper envelope that rattled with pills, and left for the bridge. I choked down my first dose of antibiotic, swallowing a tablet with the last of the Gatorade. The liquid's warm temperature and yellow color, combined with a slightly chemical taste that reflected its ingredients, made me feel like I was drinking urine, which was fitting, since this whole situation was pissing me off. I conquered the urge to throw up again and, finally, after two days at sea, felt ready to start a day's work.

As I climbed down the ladder to the lower deck I mulled over my first experience with long-distance medicine, and begin to see flaws in the system. A scenario arose in my mind:

CAPTAIN: "Chief, this TELEX is a little smudged. Does it say 'Give one ampoule of antibiotic' or 'Got to amputate at the neck'?"

I decided it might be prudent to keep all future medical complaints to myself.

I entered the main lab to find that the on-watch science team had already divided themselves into areas of responsibility. They seemed to have quickly settled into the routine at sea. Everyone was busy with something. The space was set up well for controlling instrument deployment. A semicircular central console with about 30 TV and computer monitors gave the status of every piece of equipment on board. Numbers scrolled from top to bottom of screens as data streamed into the data-logging computers at the rate of one datum per second. Another set of monitors depicted the ship positioned on an electronic chart, a red line trailing behind to show where we'd been. Video feeds to all parts of the vessel were also broadcast here, giving the technician on watch the ability to view activity on deck. If it weren't for the low ceiling and the fact that all the furniture was bolted to the deck, the place could have easily been mistaken for mission control at NASA.

One of the science technicians was standing near the chart table reading a manual. I asked him how things were going.

"Just trying to work the bugs out of the ADCP," he replied without looking at me, then remembering my recent condition, added, "feeling better?"

"Anything would be an improvement over yesterday," I said.

"Well, not much happening right now. With any luck I'll have this working soon."

Theoretically the Acoustic Doppler Current Profiler was always on, measuring the speed of the currents directly below the ship at all times. The data were useful in validating the data from other instruments as well as providing a picture of the current all along the ship's track for the entire cruise. That was the intent. It turned out that this was the first time the ADCP on this ship has been turned on in four years, because *Raven* was used mainly for cruises that studied the geology beneath the ocean floor, and geologists don't usually have much interest in currents. The ship was also equipped with instruments that gathered data on bottom and sub-bottom composition, and sensors that painted a detailed picture of what the bottom contours looked like.

Most of the technology on board was of little use to Amy and her colleagues. The only other device beside the ADCP that was of interest also did not seem to be working. A conducting cable that would be used to lower the rosette and transmit data back to the ship was not performing the data-delivery portion of its function. Amy was across the room talking to one of the technicians. She did not look happy.

"Charlie, I need to know, is the problem with the instrument we brought or with the hydro cable that the ship supplied?" she asked him. I could tell by the flushed look on her face that this conversation had been going on since before my arrival in the lab.

"I've checked our package twice and I get a signal. When I hook it up to their cable, that's when the errors start," Charlie responded, somewhat defensively.

Amy countered, "their tech insists that the problem is on our end. At this point it doesn't matter whose fault it is, we just need to fix it."

She walked back to where I was standing, out of earshot from anyone else in the lab.

"Things are pretty screwed up with the wire," she told me. "We're getting intermittent signals to the unit. We should have brought our own cable." She flicked open the crystal bezel of her Braille watch and fingered the position of the hands. "I am hoping to get our first cast done by the end of this shift, but it's not going to happen unless Charlie figures this thing out. I'm beginning to think we should have brought our own winch too." She seemed angry with herself for not anticipating a problem like this.

With nothing to offer in the way of a solution, I walked to the room containing the stacked packing cases of RAFOS floats and began to open them up. They wouldn't be needed for a few days, but given the problems Charlie was having to deal with, I wanted to make sure there were no surprises waiting for me as well. Things had not gotten off to a good start.

21

BAB AL MANDEB: the constriction at the southern end of the Red Sea where all Suez Canal traffic must pass. We darted back and forth across lanes of heavy shipping, a squirrel dodging 100,000-ton tankers on an oceanic highway. Every few miles we stopped to deploy instrumentation.

Here the Gulf of Aden was at its narrowest. High rust-colored cliffs rose to a barren moonscape on either side of us. A few islands of bare rock to the northwest formed giant stepping stones from Africa to Asia, creating a geographical limit to our exploration.

I was standing on the bridge when one of the crew spotted a small, open boat following our wake. All eyes were glued to the craft. Leslie, the third mate, kept one hand on the phone receiver in preparation to alert the captain if necessary. "What do you think?" she asked the AB on watch with her.

"I don't know. Maybe they're just riding behind to let us break through the chop," he said without dropping the binoculars from his eyes, an unlit cigarette rolling from side to side across his lips as he talked.

"We'll give him a few more minutes. Let me know if he gets any closer." Leslie kept her fingers wrapped around the phone.

I picked up a spare pair of binoculars and focused on the small boat. It looked like an old wooden sailing dhow that had been outfitted with an engine. A stump of a mast supported a fishing net which hung like a veil over the middle of the boat. Two men in white robes sat in the stern, while a shirtless man stood at the bow looking directly at me. They didn't look dangerous, but why follow us like a seagull behind a shrimp boat? It was not that rough out here that they needed us to smooth the water.

After another 10 minutes our visitors turned 90 degrees and headed east. "Thank God. I don't need this kind of shit," Leslie said as she put the phone back in its cradle.

There was no Jim or Vince on board to shout "Condition Alpha," or Gamma, or any other Greek letter. The operators of *Raven* had determined that because we experienced no trouble on the *Knorr* cruise, they would save some money by only providing basic counter-piracy training for the crew during *Raven*'s transit through the Red Sea. To me this was flawed logic, the equivalent of canceling your auto insurance because your neighbor never had an accident.

Suddenly Leslie disappeared from my view. She had been standing on a small, custom-made platform which I had not seen before, and her four-foot-five-inch frame became lost behind the bridge console. Her stature was her only diminutive characteristic. Though she looked like someone not old enough to drive a car, let alone this ship, her voice commanded authority.

"T, make a note in the log with a description of that boat in case it comes back on the next watch," she said, looking out the starboard windows.

T, or Thomas, was a Jamaican, complete with Rasta hair and "ganja-mon" accent. His green, black, and yellow knit cap was forever affixed to his head even though the temperature on deck approached triple digits.

"I bet the book says something about hazardous duty pay," he stated, and I got the impression that the comment was directed as much to me as to Leslie. Thomas was forever quoting the union rulebook. I would not have been surprised to find that he knew it better than he knew the rules for preventing collisions at sea.

Leslie lifted a pair of binoculars off her chest and looked in the direction of the receding fishing boat.

"You keep bitch'n about rules, T, and you'll never make mate."

"I'll leave you two to fight this out yourselves," I said as I finished copying the navigation information I needed and went below to the lab. I didn't bother to tell Amy about the boat. It seemed trivial after few minutes had passed, and by the time I reached the lab I was not even thinking about it.

It was late in the evening and an hour before my watch was to end. It would be at least two hours until we reached the next station. Amy was working at her computer, making preliminary plots of the data generated so far. I took the rare opportunity for free time to go on deck and watch the moon rise. The air, if not exactly comfortable, was at least noticeably cooler than in the afternoon. As the wheel of the watertight door spun closed behind me the relative quiet of the ship's interior contrasted with loud music suddenly stabbing at my ears.

A bare-chested man with a bandana around his head was dancing under a single flood light to a tuneless beat blasting from the aft deck stereo speakers. In each hand he twirled a tennis ball tied to a cord wrapped around his fingers. He looked like a luau act at the Kona Hilton. Before I could ask, he volunteered the answer I feared he would give.

"I usually do this with the balls lit on fire, but open flames aboard ship make the captain nervous." He didn't take his eyes off the spinning orbs. His name was Jerry; he was the ship's computer technician. He was young, with a skinny, taut build and a thick mat of curly black hair cascading over his shoulders.

I sat down on a chair-sized bollard and waited for him to finish. He walked over, beads of sweat dripping from the tip of his nose. "It's even more awesome when the guns are going off," he said, pointing to a web of steel pipes and rubber hoses suspended above our heads.

"Guns?" I replied, looking up into the tangled mess snaking along a rectangular metal frame.

"Air guns. These booms swing outboard from each side and we trail an array of those aluminum tubes behind the boat. We fire charges of compressed air through the hoses and when the air comes out of the tubes at the end—BANG!" His arms flew apart to emphasize the noise.

"What are you shooting at?"

"The bottom. This shit is so loud it can penetrate the bottom sediments. When the sound gets reflected back up to the ship we analyze the echo and determine the composition of the sea floor. The geologists love this stuff. The guns go off every 15 seconds, all day, all night. You should see it. Shit, it looks like those depth charge scenes in World War II movies. It's cool to watch. The stuff you're doing is pretty tame by comparison."

"That's fine with me. I don't think I could take that kind of noise reverberating through the hull . . . every 15 seconds . . . of every minute . . . of every hour . . . of every day." My stomach shuddered at the thought of having to listen to Armageddon for weeks at a time.

"You're not the only one. Greenpeace chases us around the world complaining we're making whales and dolphins neurotic with all the noise. That's why your group is here. We can't seem to fill the ship schedule with seismic work anymore, so we're trying to diversify what we do."

I couldn't tell whether he was happy for the change of routine. "So what do you think about the risk of pirates out here?" I asked, changing the subject.

He looked around the deck before replying. "*Raven*'s a small fish. It's a big pond. Ten containers of DVD players, now that's worth stealing. But—" He paused, playing with the string around his finger. "I don't know how those people think. It only takes a couple of crazy fucks to ruin your day."

I went to bed thinking about pirates, but feeling thankful for not being subjected to the roar of air guns on this cruise. My thankfulness was short-lived. It turned out that the air guns were not the only guns on board.

At 6:00 a.m., six hours before the start of my watch, I was awoken by the nautical equivalent of a rooster—the needle gun. The needle gun was probably invented about nine months after the first steel ship was launched, its primary purpose being to chip rust off of steel.

Needle guns make a lot of noise. When metal hits metal at high speed a sound is created of a most uncomfortable frequency, that resonates throughout the ship, giving one a feeling not unlike being sealed in a 55-gallon drum while someone goes at the lid with a jackhammer.

I opened the cabin port and stuck my head out to identify who was responsible for truncating my rest, only to be surprised by a familiar face.

"Calvin?" I said, uncertainly.

"Yep," he said, lifting a set of protective goggles from his eyes.

Calvin hadn't lost an ounce since we'd sailed together on *Knorr*. If anything he'd gained more weight. I continued talking, thinking that it would eat into his rust-busting time. "Remember me? On *Knorr* last March?" He squinted at my face, framed by the porthole, trying to form a recollection. I didn't wait for him to answer. "What are you doing on *Raven*?"

"I'm filling in here as oiler for a couple of months, till I get my bail bondsman's license down in Kentucky. Sure won't miss going to sea, and going after assholes who jump bond is no worse than working in the middle of all these Arabs."

It was the way he said Arabs—long "A," space, "rabs"—that made me think he was among the growing number of *Raven*'s crew who weren't happy to be here. Unless morale improved it was going to be a long cruise.

I convinced Calvin to direct his weapon's fury somewhere further aft and went back to bed only to have a nightmare about fat men in skimpy tights twirling flaming balls around on a string to protect the ship from pirates. I woke up from this dream, surprised to find myself wishing that Jim and Vince were aboard. Despite their humorless intensity, at least they provided us with some sense of security, even if only an illusionary one.

22

WITH THE EXCEPTION OF "Needle Gunner – First Class" Cal, the bridge was the only place so far where I had interacted with *Raven*'s crew. In the scientific spaces crewmen seemed as scarce as a cold beer. If a ship's electrician or deckhand passed through the lab he or she tended to move quietly along the wall like a mouse evading a cat. One such crewmember entered the room that I'd been allocated to set up the drifting floats. He was a big man, huge really, the size of two men. He tried to sneak out without saying anything and I verbally pounced on my oversized prey.

"Hi, what's up," I asked before he could open the door at the other end of the room.

"Naw much. Check'n tank levels," he said, holding up a coiled metal tape with a weight attached to the end. He stood there looking uncomfortable, and judging by the sweat soaking through the scarf tied tightly around his bald scalp, he had just come up from the engine room.

I tilted my head back and introduced myself. His name was Bear, though it was doubtful that was the name his mother gave him. The only other unimaginative moniker for him would have been "Tiny." He put a greasy rag in the pocket of his overalls and shook my hand with his paw. It seemed we had the two largest oilers in the Merchant Marine serving on the same ship. If this was some kind of omen, I couldn't decide if it was a good or a bad one.

"You guys picked a nasty place to work," he said.

"Yeah it's pretty hot out."

"Naw, I don't mean that. I'm from Louisiana, I can take the heat. Pirates is what I'm talking about," he said, dabbing his forehead with the greasy rag.

I asked him more questions and we had a more or less one-sided chat for a few more minutes before Jan, a young graduate student, walked in. The contrast between her and Bear could not have been greater. Every characteristic of one was the complete opposite of the other. Bear's flesh had the color of walnut wood and seemed to be as hard, while Jan's was the translucent white of rice paper. She stood as tall as his waist. She didn't seem conscious of the contrasts between them when she spoke.

"Amy asked me to bring these down to you." She giggled as she handed me a stack of papers.

Even their speech was from different worlds. Jan did not so much talk as laugh out her words in a nonstop, nervous flutter of vowels. When Bear spoke at all it was in a frequency so low that the words were more felt than heard.

It was in this deep bass that Bear excused himself and squeezed through the compartment doorway. Talking to more than one stranger at a time would probably have taxed all of his social skills.

"Thanks," I said to Jan.

"No problem." Another laugh, and she floated through the door a few steps behind Bear, leaving me alone again with my floats.

I looked at the paper; it was a revised list of RAFOS float deployment sites. I went to make the necessary corrections on the chart.

Raven steamed on, continuing our mission to repeat the same 238 stations we had completed on the previous cruise.

Despite the windy day the temperature remained constant—hot, or in the appropriate words of Neil Simon, "Africa Hot." Out on deck, preparing to deploy the first set of RAFOS floats, I felt like an ant under a sunlit magnifying glass. It was breezy and there was a uniformly thick haze over the Gulf, which limited visibility.

The floats were not to be released at every station, but at selected sites where Amy thought they would stand the best chance of being captured in a subsurface current of Red Sea water. Based on part science, part art, and part instinct, the decision about the optimum place to launch the precious few floats we had was never an easy one, and Amy spent hours poring over charts before marking the stations. In this part of the job Amy's colleagues were as blind as she was when trying to picture what was happening thousands of feet below us.

The RAFOS floats looked like six-foot-long by five-inch-diameter glass test tubes. At one-quarter of an inch thick, the glass provided a housing that was relatively inexpensive, easy to manufacture, and capable of withstanding the tremendous pressure of the deep ocean. An end cap sealed the open end of the tube. Besides keeping out water, this cap provided a mounting surface for the ballast weight and hydrophone, a sensitive underwater microphone designed to hear the tone emitted by the sound sources.

These floats had become my little pets. Each one was an individual with its own serial number and quirks of construction. As I sent instructions from my laptop through the wire umbilical and into a float's memory, I wondered what it would endure stumbling in the darkness of such crushing depth, while moving at the whim of the current.

Searching an open crate under the bench for the weight matched to a particular float, I found the one with the correct serial number stenciled on it. After double-checking the number, I secured it to the end cap by a monofilament lanyard. Matching the weight to the float was critical, as each one was ground to within a gram of the desired ballast.

Water is not uniform in density from the sea surface to the sea floor. Temperature, salinity, and pressure all combine to give seawater its density. This difference in density allowed Amy to match the buoyancy of the floats to the depth of the water she wanted to study. When a float sank to a water density where it was neutrally buoyant, it would not sink any further, but would instead drift at that depth until the weight of the float changed. For the RAFOS float this change would come when the electronics inside signaled for the ballast weight to be dropped. This sudden change in buoyancy would cause the float to pop to the surface. Bobbing only a foot or two above the wave tops, a foil antenna visible through the glass housing would broadcast to a satellite passing overhead all of the data collected during the float's one-year mission. Once the data were transmitted, the float would drift in the surface current until it sank from leakage, the weight of marine growth, or collision with a ship. Some floats had been captured in fishing nets, while others washed up on distant shores. The finders of such prizes dreamed of large cash rewards, but the reality was that the expense of attempting to retrieve an expended float usually exceeded the $5,500 cost of a new one.

As I stood over the lab bench, eight slender tubes reflected the florescent lights above, bending the image of the room around their shiny glass surfaces. My image was visible in each one, body distorted, shrunken head sitting atop a bloated belly. Looking through the glass at the green circuit boards dotted with numerous electronic components, I wondered for the 10th time that day if the instructions I programmed inside those little ceramic boxes were the right ones. If a battery died prematurely or an internal clock stopped the float would fail, a failure that clearly rested on the shoulders of the designer, manufacturer, or assembler and not on me, the lowly technician. But with almost no chance of ever recovering a failed float, the question would always remain: who screwed up? It was like trying to solve a murder without a body. There was lots of room for finger-pointing and inevitably, some of the fingers would point at me.

Amy joined me in the lab and we carefully lifted the first float out of its cradle and maneuvered it onto the aft deck.

"Let's only take out one at a time. I don't want them to sit in the hot sun any longer than they have to," Amy said.

Amy used the tail end of the float as a guide while we slowly wended our way through the narrow passage. I led the way, sensitive to any obstructions lying in wait to take a bite out of the fragile glass float. Stepping through the last bulkhead and onto the exposed aft deck, I felt like someone had opened the door to a blast furnace. The heat of the steel plate penetrated the soles of my shoes.

Deployment of these instruments required three people, one of whom had to lie with his belly on the deck at the ship's fantail. With the rest of my watch looking at me, I eased myself down on the near-molten metal. The life vest I was required to wear offered some insulation, but rode too far above my waist to be totally effective. You've heard the expression "it's so hot you can fry an egg on it." Well, somehow the only image that crept into my mind as the flesh on my elbows blistered was of my family jewels, now firmly in contact with the steel ship, bursting like corn kernels in hot oil.

Leaning out as far as my arms would allow, I supported the lower end of the float while Amy guided the rest down from above. The two of us looked like we were inserting a six-foot rectal thermometer up the ship's stern

"Have them put a couple of turns on the screws and then stop," Amy called over her shoulder to a deck hand holding the intercom to the bridge. Her intention was to provide just enough way on the ship to allow the prop-wash to push the float from the hull and propellers as it sank. Though it was hard to imagine a scenario where the float would drift back under the ship after it was dropped, Amy took no chances.

Turbulent water began to stream back from under the hull.

"All stop," said the deck hand.

With that, Amy and I let the float slide between our fingers. I watched for a few seconds as it descended silently into the clear blue water. It quickly disappeared from view, a shimmering and twisted glint of light turning fainter with each meter of depth.

After deploying four floats at this station I stood up drenched in sweat. My forearms were red from leaning on the deck but there was some sense of hope that my ability to procreate remained intact. We retreated to the more temperate climate of the lab and worked inside until the night brought respite from the sun's heat.

23

SHOWERED, SHAVED, AND WITH AN HOUR UNTIL THE START OF WATCH, I could focus on nothing but a big breakfast of pancakes and ham as I strode purposefully down a long corridor toward the galley. In the main companionway Jan was hurrying in the other direction.

"There are pirates circling the ship," she blurted out with a giddy excitement that defied the seriousness of her words.

"I've got my Shellback card in my cabin, go tell it to another sucker," I replied, with the hoarseness of sleep still in my voice.

"I'm serious, everyone is to muster in the lab," she said, squeezing past me at the top of a narrow stairwell. Practical joking went on every day aboard research ships. I decided not to bite at this one and slowly dropped from step to step, thinking only about something to eat.

At the bottom of the stairs I was almost knocked over by the steward. He too was smiling.

"I'm on my way to my post to stand guard over the refrigerator," he said in mock formality while raising his hand in salute. Maybe there *was* something serious going on.

"What's happening?" I called up after him, my voice echoing off the walls of the narrow stairwell.

"A boatload of armed men is circling the ship," he yelled back without turning around.

The closest porthole was in the lounge. I pulled back the ever-closed curtain. Framed like a photograph in the round bronze bezel of the glass window was an open boat with six men aboard. *Raven* was stationary and the skiff that circled maintained a separation of about 30 meters. It was about 18 feet long, of modern fiberglass construction, and propelled by a high-horsepower outboard engine that looked equally new.

What seemed incongruous to me was not the fact that this little boat filled with men was so far from shore, but that one of them was waiving an AK-47 automatic rifle in the air, its distinctive silhouette dark against the bright sky. With the backlighting from the sunlight, it was hard to tell if the men aboard the skiff were wearing uniforms, though in this region uniforms were not necessarily an indicator of formal military association. A man at the stern holding the tiller of the outboard motor pointed to the others in the skiff as if giving instructions. With the exception of the one holding the rifle and shouting at us, the faces of the sextet remained passive. The scene was so surreal and unexpected that it didn't provoke the sense of fear in me I would have expected, but rather a strange excitement, like I'd witnessed in Jan and the steward a few moments before. All of our concerns about piracy were materializing, and our reactions were at odds with the seriousness of the situation. This would soon change.

The disparate size of the two craft made the threat seem impotent. The little boat looked like a flea hoping around a dog in search of a good place to bite. But when the man with the rifle momentarily raised the gun's stalk to his shoulder and took aim at something or someone aboard *Raven*, things suddenly seemed very real. My thoughts turned immediately to Amy and I quickly exited the lounge in search of her.

I opened the lab door. Cold air swirled around my ankles. Most of the science party stood in stiff poses around the VHF radio and closed-circuit TV monitors. The small boat came into view on one of the screens as it passed within the field of the starboard camera. Amy leaned with her back against a bench, arms crossed, her face tense but otherwise betraying no emotion. I walked over and leaned against the same bench, close enough so that the bare skin on my arm touched hers. She pushed back against my arm. Her flesh was cool and even through this small bridge the strain in her body came across. "Have they done anything yet?" I asked.

"Nothing but circle the ship and yell a lot in Arabic," she replied.

"The rosette?"

"Coming up now. We can't move until it's on deck."

The instrument was in the water. A glance at the screen showed it at 200 meters and coming up fast. I could hear the whirr of the winch on the deck above pulling wire in at the maximum rate. Another camera showed two people outside, ready to secure the rosette as soon as it came aboard.

"How are you doing?" I asked quietly.

"OK. We'll see what happens when we start to move." Amy stood with her eyes closed, assessing the situation based on what she could hear.

The captain's voice suddenly filled the room from the VHF radio speaker. "Chief, let me know when the package is secured on deck. Have the hoses charged at the stern on both sides."

Not satisfied with watching a detached and grainy video image, I walked toward the lab doors that led out on deck.

"I'd rather you didn't go out there," Amy said, not as the chief scientist giving an order, but as a concerned wife.

"I'll be fine, I'm just going to get a better look."

"Be careful."

"The captain instructed anyone on deck to ignore them," said Jan as I spun the wheel on the watertight door. How do you ignore someone pointing a gun at you, I thought, pushing open the heavy door.

The first inhalation of hot air scalded my lungs. I'd stepped on deck just in time to see the rosette lowered none too gently into its cradle. I almost lost my balance as the ship lurched forward, accelerating as quickly as possible away from the pirates. Black smoke spewed from the stack, indicating that the throttles were pushed to the stops. The skiff came back into view near the stern on our starboard side, circling us effortlessly. It was as if we were not moving at all. I found myself willing the ship to go faster.

The small boat buzzed past again. One of the men in it walked forward with something slung under his arm. As he reached the bow he held the long cylindrical object over his head, and then raised and lowered it repeatedly like a weightlifter pushing a barbell. It was an act of defiance clearly meant to convey the message, "You can't run from this." My knowledge of weaponry came from war movies and CNN, but it was obvious that the device this man pumped in the air was a rocket-propelled grenade (RPG) launcher. I didn't see a grenade attached to the end of the launcher, but had to assume a cache existed within reach.

There were two types of grenades that could be fired out of a shoulder-held RPG: anti-armor and anti-personnel. The first could easily penetrate the hull of a ship; the second, if lobbed through a bridge window, could kill or severely injure everyone in that space. Regardless of which type these men had, the introduction of the RPG made the situation much more serious.

As the skiff was lost from view around the bow, word was passed of the captain's order for all nonessential personnel to muster in the main lab. The only members of the crew who were to remain on deck were those designated to man fire hoses and repel boarders.

I stepped back into the lab. The sweat on my back turned to ice water in the air-conditioned room. Everyone clustered around the TV monitors while Jerry pushed the deck camera selection button repeatedly. Images of various areas of *Raven*'s decks flashed on the small screen until a camera was found that showed some activity. The picture was grainy, but legible. One of *Raven*'s crew, fire hose under his arm, stood along the port stern rail looking out at the passing skiff. From between the tight pack of bodies in the small boat another rifle was raised, and its barrel swung toward *Raven*. Suddenly the crewman we were watching on the monitor dropped his hose and fell prone on the deck. A warm rush of adrenaline surged through my body as the knowledge of what I had witnessed assembled in my brain.

"Shots have been fired. Shots have been fired," the captain's voice said over the PA system. His tone was disturbingly casual. Our eyes were glued to the seaman now scampering toward the nearest door. He appeared to be unhurt. A chorus of exhalations filled the lab, but nobody said anything. There was nothing to say, as everyone tried to assimilate this new level of threat.

"Well, let's get everyone inside and secure all watertight doors. It looks like they took some shots at us," the captain continued a few seconds later, tone unchanged. A knot was building in my stomach. "All personnel not assigned to a security detail please return to their cabins and lock their doors." With that announcement, a surprisingly silent line formed as we exited the lab and dispersed to our cabins.

I locked our door behind us and hurried across the room to the large window on the port side. Moving back the curtain, I could see the white skiff, still in pursuit, about 30 meters off our port quarter. I described all this to Amy, who had taken a seat on the couch. She was leaning forward, elbows on her knees and head in her hands. "What do you see?"

"Two men are sitting on the gunwales, the others are lower in the boat. They probably don't like getting wet. It's pretty choppy out there."

It looked like an uncomfortable ride. The skiff drifted further aft, almost out of view. I pressed my cheek to the glass, straining to see what was happening. A couple of the men repeatedly raised arms and rifles in the air, and though their voices were muted by the hull, the aggressive body language left no doubt that they were still intent on engaging us.

Amy suddenly stood up and grabbed the handheld VHF radio off the desk. She turned it on and we could hear the bridge communications among the crew. The captain was asking for updates from crewmembers stationed aft. I lost sight of the skiff and turned on the close circuit TV in our room to glimpse what we could from the video feed of the deck cameras. It wasn't much but it provided some sense of control, if only an imaginary one.

We both sat in silence. The fuzzy picture of the skiff still keeping station on our quarter seemed less threatening in its lack of detail, but the situation hadn't changed. With the freeboard of *Knorr's* transom only a few feet above the water, it would be a simple task for those men to board our ship.

A look around the room reinforced my feeling of helplessness. What if they did come aboard? Should we try to stay hidden? There might have been hundreds of places to hide on the ship, but to what end? If someone came through the thin cabin door, what would we do? Fight them? My mind raced with the possibilities. I could sense Amy thinking through the same scenario, but she offered no plan either.

The video image on the TV was useless; the camera was too far away and didn't provide enough detail. I could get a much clearer picture by looking out the porthole. When I did so, I saw that the skiff was back in my field of view.

"No change," I reported to Amy. "They're still at about the same place. I don't get it. They can climb aboard anytime they want. They could have easily done it when we were on station."

Amy leaned back on the couch. What I saw in her face wasn't fear as much as concern. We were both scared for ourselves, but she felt the additional burden of responsibility for the other 43 lives aboard, justified or not. She was the reason we were here, and I could tell she was reminding herself of that fact with every passing second.

Since Amy couldn't see anything of consequence out the window, my job was to maintain a running commentary. I gave her the facts and let her evaluate them. At times the skiff moved closer and then backed off again. Periodically, arguments seemed to erupt among the six men, though their in-fighting was apparently not enough to distract them from the task at hand, and the chase continued unabated.

After 15 minutes, the small boat slowed to a stop and I breathed for what seemed like the first time in an hour. "They're stopping!" I yelled, trying to sound hopeful. "It looks like they may be giving up."

Before the words left my mouth I knew I had spoken too soon. One of the men pulled up the RPG launcher. It was handed quickly from man to man until it reached the skinny figure sitting on the bow. As he slung the four-foot tube onto his shoulder, it became clear that they'd only stopped the boat for the purpose of aiming the weapon. I wanted to tell Amy what was happening, but before my brain could move this new information to the area responsible for speech, a puff of smoke and flame spit from the back end of the launcher. Three splashes, like those of an invisible stone skipping across the water, progressed toward us. A millisecond later a large orange fireball erupted a few meters from *Raven*'s port quarter.

"They just fired a grenade at us!" My voice was calm, but inside my head another voice was screaming, "You idiot! How smart was that to stick your face in a window and look down the barrel of a bazooka?"

I sat back down on the couch with Amy. The cabin that had seemed so big when I'd come aboard now felt claustrophobically small. Amy's eyes were red and she was holding back tears. "What if someone gets hurt?" she said.

What was there to say? I tried not to think about the possibility of someone being killed, and focused on the positive. At least Amy and I were together and, for better or worse, knew what was going on. Sequestered in even smaller cabins below the water line, the others had no portholes or radios or video images, no idea of what was happening.

I forced myself to breathe slowly and remember that as long as the ship was separated from the men in the boat by even a small expanse of water, things would be OK.

"We're protected from bullets or grenades by the steel hull of the ship," I said to Amy, secretly hoping that the grenades they had couldn't penetrate the hull. With this thought came the realization that if those men were willing to lob explosives on deck and risk killing someone, they'd be willing to do just about anything if they came aboard.

Raven's course was a little east of north as the captain tried to put as much space between the coast of Somalia and us as he could. There was nothing to indicate that this maneuver would act as a deterrent, but it seemed logical to run away from the source of our problem, and it was a reasonable assumption, based on where we'd first encountered the skiff, that these men were Somali.

Sitting still on the couch became impossible, so I moved once again to the window. More gesticulating aboard the skiff. Hands were flying in the air, this time at each other and not at us. There did not appear to be one person in charge, and no radio antenna was visible, so it was not likely they were in communication with another boat. The wind was pushing up larger waves now, making the ride even rougher for the men in the open boat.

"Hey! They're slowing down again!" I called out to Amy.

"Their fuel supply is limited. They'll have to act soon or give up," she said. While I had been focused on the action, Amy had been calculating in her head how far we had been traveling.

As soon as the boat settled down on its own wake, the man in front raised the RPG launcher again, and any hopes I had of them turning tail vanished.

This time, not waiting to see if the marksman had improved his aim, I ducked down behind the protection of our steel cocoon. Nothing. No explosions, no shattered glass or ringing metal.

"What's happening?" Amy asked impatiently from the couch.

Sticking my head up, I watched as the skiff started up again. It came as close as it ever had. I could now see the details of each man's face. They looked no different from the merchants we'd encountered in Djibouti. One of them could have been the bar tender at the hotel, another the bellman. None had the face I expected to see on a pirate.

The men had begun to wave their hands at each other instead of us. "It looks like they can't decide what to do," I said.

All of a sudden, the bow of the skiff turned abruptly to port, away from *Raven*, and the boat headed back through the white foamy gash of *Raven*'s wake.

"They've turned away! They're taking off!" I shouted.

Amy's head fell back on the cushion and she stared up at the ceiling, all the muscles of her body sagging in relief.

Within a few minutes the captain's voice came over the PA system. "The pirates have stopped pursuing us. Everyone can come out of their staterooms." We opened our door and walked to the deck below. There we found others coming out of their cabins, wide eyed and dazed-looking. Low tentative murmurs gave way to giddy staccato talk and laughter as the adrenalin, with nothing to act against, sought an alternate route from the body.

24

WITHIN MINUTES OF THE ATTACK'S END, satellite phone calls and FAX messages began bouncing around the globe. Amy started negotiating with the powers that be to let her continue work and not force her to abandon the project. Two hours later we got word that officials at the State Department, The National Science Foundation, the Eastern Ocean Research Institute where *Raven* was based, and the Woods Hole Oceanographic Institution had agreed to let us continue working as long as we approached no closer than 50 miles from the coast of any Gulf country. Since the Gulf was only about 150 miles wide, this set a significant limit on our area of research and one that could have ended the cruise. But Amy had presented a successful case to the string of decision-making institutions by arguing that she could still get useful data despite this radical change and the short amount of time available to re-plan the entire cruise track.

She took up the task of re-planning the remaining stations, including all the necessary time and distance calculations, with an energy and agreeability that swept away any grumbles from the science team about the restrictions making their work impossible.

This ability to confront a problem and see it through to the end was what had attracted me to Amy from the beginning. It was the opposite of what had driven me away from my first wife. Where she would gravitate toward ignorance of difficult problems, Amy would take all the shit thrown at her and deal with it. Maybe not without complaint, but hey, blind scientists who are shot at in the middle of the ocean are allowed some slack.

I redrew station lines on the lab's chart using the new 50-mile limits. The resulting cruise track looked odd when compared to the old one. The long zigzagging lines that before had seemed to nick each shoreline now oscillated around the middle of the Gulf in short staccato strokes.

"I'm confident that some floats will drift closer to shore, into areas we can't go," Amy said optimistically. "The problem is, that is the area I'm most interested in. At least most of the sound sources were put in place on the last cruise." I also couldn't help wondering if there was a slight ironic sentiment behind her words. The likelihood that some of those floats could eventually drift beneath the boat of our recent attackers was not beyond the realm of possibility.

I tapped the chart on a point outside the new boundary line off the southeast coast of Yemen. "Yeah, but the only other mooring we have to put in goes here." I measured off the distance from the coast and gave it to Amy. It was 35 miles.

"I'll have to negotiate that with the captain when we get closer. Maybe people's nerves will have settled down by then," Amy said.

My eyes drifted to the last sound source mooring position on the chart. It was not that far from the port of Aden in Yemen where the U.S.S. *Cole* had been bombed the previous year. We really were sailing in the jaws of a wolf.

The pirate attack was behind us, but by the next day an air of uneasiness was to be felt among the crew. The talk in the mess had a mutinous undertone, as previous dissenters felt vindicated by their prediction of trouble. Some said we should pack up and go home. Of the entire crew, T was the most vocal.

"Man we're like prisoners on this ship. No getting off until the warden on the bridge says so," he said to me as he sat down at the table. He and Bear had cornered me on the mess deck.

"Did ya see the latest news about dem Italian dudes on that

fishing boat?" Bear asked me while pointing a thumb over his shoulder at the ship's bulletin board. I had seen it. Someone had been posting email news messages related to pirate attacks since we'd left Djibouti. The captain was not happy about it and took them down whenever he walked by. The latest news was of eight Kenyan men from an Italian-registered fishing boat held hostage in Somalia. The most recent ransom demand was one million dollars.

"They've no right to keep us out here after what we went through. It's in the rules." T rapped on the table with his spoon for emphasis.

"Hey, I'm hired help just like you. I don't say where the ship goes. You should be talking to the captain," I said defensively.

"But your wife can call this whole thing off. The captain ain't going to stick around if she says she's done work'n," said Bear.

It was obvious to me that T had put him up to this, and not wanting the role of middleman in their version of shuttle diplomacy, I tried to remain noncommittal. "It's clear neither of you have ever worked for your wives. All I can do is tell her that you're not happy. I can't promise that it will change anything." I was not lying. I would tell Amy how they felt, but I was not going to be their advocate.

"The pirates gave up. We won," I told them. "Besides, it's like lightning, what's the chance it will strike twice?"

"Oh yeah, and what if those dudes in the little boat come back with more dudes and a bigger boat?" T said.

"Or bigger guns?" added Bear.

"We're staying well offshore. I doubt they could even find us again," I countered, but I didn't think my words had any effect. I finished eating and went on deck, not so much for the cool night air as to avoid having to listen to T, and to Bear, who suddenly seemed to have found his voice.

Sitting on a large winch drum, I watched Jerry cast his line repeatedly for cuttlefish. They darted in and out of view, drawn to the oasis of light that surrounded the ship. Their pale orange bodies flashed iridescent colors as they neared the surface.

"What do you plan to do if you catch one?" I asked him.

"I don't expect I'll catch one, but I think you can eat them," he said.

A fish by name, the cuttlefish was more closely related to the squid in the way it looked and swam. The closest most people came to a cuttlefish was a parrot cage. Parrots like to sharpen their beaks on the unique spear-tip-shaped bones of these fish.

Looking up from the pelagic ballet below I saw the lights of a ship passing less than a mile away. Jerry noticed it at the same time. "They're lit up like a Christmas tree. Bet their fire hoses are charged and pumping over the side."

"I hope they have more than bright lights and a garden hose," I added. We both watched the ship grow smaller as it steamed west toward Bab al Mandeb. Since the pirate attack I had felt like a seasoned warrior, but I suspected all the fear would resurface if we encountered trouble again. Judging by the way Jerry was looking at the freighter, he probably felt the same.

The redrawn track lines on the chart were significantly shorter since our confinement to the new sailing area limits. The old lines remained as faint traces despite judicious rubbing with a pencil eraser, and it was anybody's guess what effect this constriction in scope would have on the resulting data. Amy did not seem to be deterred, however, and remained encouraged by what she was seeing so far.

"So what's the scuttlebutt? Is there still talk of jumping ship?" she quizzed me over dinner.

"Some people continue to grumble, but we've come this far without T managing to lead a revolt. Anyway, he just likes to complain. If it wasn't about this it would have been about the food, or the work, or the heat."

"I'll be happy when we're done. I'm ready to go home," Amy said, sounding more tired than she looked.

"Me too. If this was my first cruise, I don't think you could get me on another," I answered through a mouthful of mashed potatoes.

25

IT HAD BEEN A WEEK SINCE THE ATTACK, a week that saw a return to the routine of shipboard life. In another 24 hours we would be back in Djibouti. With the bulk of our work finished and the end of scheduled watches, Amy and I tried to pack our belongings. I was searching about our cabin looking for any personal articles that had come adrift since our arrival on board. A pocketknife reappeared under the couch and an unread paperback book got tossed into a duffle. Bills and loose change in various world currencies had found their way to the back of my bunk-side drawer. The need for money felt so remote aboard ship that it got quickly forgotten, left to settle under important commodities like packets of motion sickness pills and candy bars.

The thought of going home had never been more welcome, and it was heartening to perform such mundane activities as shoving dirty tee shirts into a laundry bag, knowing that mundane task was bringing me one step closer to Cape Cod. Since the pirate attack, I had not allowed myself to dwell on the return trip for fear that time would drag too slowly. With Djibouti just over the horizon I yielded to the pleasure of this anticipation.

There was a knock on our door. It was the captain. He stood there, a lanky arm holding up the doorframe, wearing an uncharacteristic downward turn at the corners of his mouth.

In his other hand he clutched a piece of ragged-edged teletype paper. He read the message in a low voice. "Two planes crashed into the twin towers of the World Trade Center. Original belief of air traffic control error has given way to suspected terrorist attack."

He extended the paper and I took it from him. It was as if I needed to see the words in print to believe them.

Disbelief dissolved into shock and then a cold reality emerged. We were halfway around the world surrounded by volatile nations, and had only sporadic contact with home. The all-too-familiar sensation of dread was beginning to spread in my stomach, reminding me of our recent encounter.

"What are your plans?" Amy asked the captain.

"I don't know. We'll have to wait and see. In the meantime we'll continue slowly toward Djibouti. I want to gather everyone in the lounge. I'll make the announcement to the ship's company then." We followed him in lockstep down the hallway.

Because of the pirate attack, people in the United States were concerned for our health and safety. Now it was our turn to worry about them. Our minds raced, trying to process what little information we had. What was happening in America?

As the ramifications of the news sank in, we realized that Americans everywhere might be at risk, and returning to Djibouti could be risky for the ship. There was a distinct possibility that we would be kept offshore for an indefinite period of time. *Raven* suddenly felt a lot smaller than it had only moments ago. Amy and I entered the ship's lounge with the captain. The small space had only enough seats for a dozen bodies. Those without a seat took up positions on the floor, knees pulled up to their chests.

The captain repeated to everyone what he'd told Amy and me a few moments ago. There was a dull silence as if the walls were papered with tufted cotton. The only sound was that of the ventilation system pushing cool air into the cramped space. What was there to say? What questions could be asked? At that point, everything we knew was contained in the two lines of the telex. Some people started to weep quietly. I scanned the faces around me, and realized for the first time that this news was even more ominous for the crew, as most of them hailed from *Raven*'s home port of New York City. These people had friends and families in the heart of the tragedy and all they had to turn over in their minds was a cryptic message delivered to them halfway around the globe. I was grateful that Amy and I were together and not separated by continents and oceans.

A low murmuring began and a shortwave radio was switched on in the back of the lounge. Even though it could easily be heard around the room, people began to move closer to the radio as if proximity to its speaker would bring them closer to home. A seaman turned the dial in search of an English language news broadcast. Within a few seconds we were straining to hear a BBC radio report through a fog of static.

A British-accented reporter was in the middle of describing the destruction of the World Trade Center towers when he broke off to announce that another plane had crashed, this time into the Pentagon.

I suddenly felt thrust into a dream. My eyes seemed detached from my body, looking down on the room below. The scene of our silent group huddled around a small radio straining for information reminded me of *War of the Worlds*, and there was a fleeting hope that perhaps this was not real after all. Amy leaned tightly against me. I did not know if the shudders in my arm were hers or mine.

After an hour, the information started to repeat itself. Many people stuck close to the radio, our only link to the rest of the world, but others had begun to drift back to their cabins to try and make some sense of the senseless. I walked with Amy to the captain's cabin. Even with the limited information she had, the science party would look to her for answers to all the unspoken questions. If that was not enough for Amy to think about, she and the captain must develop a contingency plan for our arrival in Djibouti and make travel arrangements for the science party in the event that the port was closed or determined unsafe.

The captain sat at a small desk. Amy and I squeezed into the cramped quarters. There was barely enough room for the three of us to fit comfortably. "I just got off the satellite phone with the home office. They have nothing new to tell us that we have not already heard. My orders are to sit tight for now while they try and find out the situation in Djibouti."

"What is the status of the science party coming aboard in Djibouti?" Amy asked.

"There is nothing flying out of the U.S. at the moment. All private and commercial aircraft are grounded. Some personnel may already be in transit to Europe or Africa. The equipment shipment is probably on the dock in Djibouti, but I don't know if we're going to be allowed to go in and get it."

"People have been asking me about phoning home. What can I tell them?"

"If anyone feels an urgent need to contact home the office has agreed to pay the satellite phone charges. They can use the phone in the radio room on the bridge," the captain answered.

Two hours before I was dying to get off the ship and Djibouti was a place to look forward to. Now *Raven* seemed a safe haven, an island protected by its isolation. The piracy threat of the countries that surrounded us was now magnified, and knowing so little about what was happening beyond our horizon fueled our apprehension.

For the second time that month, we floated with a sense of helplessness, a lack of control over our own lives. It was also the second time that month that Amy had to rise to the occasion and bear the burden of leadership by seeing to the well-being of the science group. Most of the time I saw Amy as my wife who happened to be the chief scientist. Now I saw something else in her. She was a professional dealing with a situation that graduate school had never prepared her for. I was proud, and thankful for the opportunity to see her work through this situation. When Amy was first diagnosed with a degenerative retinal disease, the doctor who broke the news to her bluntly stated that because of her vision loss she would have no chance of success as an oceanographer. At that moment I wished he was there with us to see what she had achieved. We'd started this trip with a sense of adventure that came, in part, from not knowing where it would lead. The only thing we felt now was vulnerability, and fear of the unknown before us, but because of Amy's infectious optimism I held on to the belief that everything would work out.

People moved around all evening in a daze. Dinner was subdued, though the relief of some of the crew was palpable after a phone call home to find out that their immediate families were safe. So far no one had reported having family or close friends injured in the attack. On the way to our cabin, we ran into the captain coming down from the bridge.

"The office has told me that they have been in contact with the U.S. Embassy in Djibouti. They indicate the risk of travel within the city has not changed since our departure. Port security, however, is questionable. The office wants me to anchor the ship in the outer harbor and use a launch to transport offgoing personnel."

At this point I was willing to swim if it meant getting home faster.

"What are you going to do about the next leg?" Amy asked.

"After we drop you off we'll probably head right back out and on to the Seychelles. It looks like the next leg is being scrubbed. The Seychelles will be a quiet place to sit things out and get fuel and provisions."

We approached the port of Djibouti on the morning of September 12. Standing at *Raven*'s bow, I watched the harbor begin to take shape, and I could see shipping activity moving in and out of the anchorage. Things looked the same as when we left, except that there was now a French naval frigate stationed at the harbor entrance. As *Raven* glided slowly past the warship, the frigate's forward cannon pointed directly at us, its turret turning to keep the big gun's sights trained on our hull. Did they know who we were? Given the chaos of the last 12 hours I could assume nothing.

There were no men to be seen on the deck of the French ship, making the motion of its guns all the more ominous. There must have been a lot of trigger-happy governments intent on protecting their borders, and I did not want to become another footnote in the day's almanac. It was too soon since the pirate attack to be looking down the barrel of another gun.

Once anchored, we waited for clearance as Djibouti port officials fabricated one excuse after another as to why they could not come out to the ship. Those of us getting off had nothing to do but look at our watches and wonder if we would make our flight out of there. Ten hours later port officers arrived with fresh ink on their stamp pads and we were cleared to go ashore.

On deck, a gentle breeze blew the smells of the port out to us and a thick, humid blanket of air had replaced the blistering heat of the day. It was late at night and from out of the darkness came a large seagoing tug. She glided up alongside *Raven*, her protective armor of black tires hanging from the gunwales. The wind had picked up slightly, generating a chop in the harbor that caused the two vessels to rise and fall out of synch with one another. Crewmen from *Raven* had placed two large fenders between the boats, preventing the grinding of metal, but the gap that they created meant it would take a tricky leap to go from one boat to the other.

Luggage was tossed over first, and then, one by one, swapping a communal lifejacket, we scampered across the divide. Spotlights illuminated the decks, but the water below was shadowed in black. "Are you ready?" I asked Amy. She didn't answer. Nor did she hesitate before climbing up on the rail.

I offered her my hand for balance. There were people on the other boat to catch her as she walked across, but with her poor vision in this light, it was a leap of faith. Amy stepped across to the other boat with the composure of someone stepping off a curb.

With everyone finally aboard we waived to the crew remaining on *Raven*. I had never met a stranger crew, but strange or not, our common experiences over the last month made for a certain bond. Those of us departing were genuinely concerned for the welfare of those we left behind. None of us knew what the next day would bring. As *Raven*'s lights faded to black I looked toward the dock more than a mile away and wondered if it would not have been smarter to stay aboard. Maybe lying low in the Seychelles was the best option. From our recent visit last winter the Seychelles seemed familiar. What was unknown now was what things would be like back home.

The Arabic crew of the tug was noncommunicative. They spent the 15-minute ride in the confines of the wheelhouse or the galley. The rest of us choked on diesel fumes as we watched the dimly lit quay materialize out of the darkness.

Our gear was piled into a waiting van and we drove to the port administration building—the same building that had been the source of so much frustration prior to our departure. Our agent's representative gathered an armful of documents and disappeared into the decaying cement blockhouse.

At the base of the quay it was hotter and more humid than it was at the anchorage. It was even more so in the non-air-conditioned van. A few of us got out into the still, night air. The taxi drivers who'd provided us with so much entertainment a month ago were gone, replaced by two dozen men in white robes milling about. As we waited a number of them come over to us.

The men formed a semicircle around those of us leaning against the van. Dressed in identical white robes and white knit skull caps, they were difficult to distinguish from one another. Their faces also shared a resemblance, coffee-colored cheeks all deeply lined from the sun. It was perhaps this fact that helped perpetuate our American bias of Middle-Eastern culture. When one sees thousands of redundantly cloaked heads walking the streets of a city, it is hard to find empathy for an individual. But as those men began speaking individual personalities become apparent.

"Very bad what happen in New York," said one man in passable English.

"Now you know how it feels," another man said, pointing a finger at me.

A third man began talking rapidly in Arabic as the first man translated. "He says he is happy about what happened. It is because the U.S. is arrogant."

"You know, most Americans don't hate Muslims," I said to the one who understood English.

"And most Muslims don't hate Americans. It is your government that they fight."

In the yellow light of the street lamps, I saw his eyes looking for my response. How could I explain that our government is a government "of the people and by the people," while still separating myself from it? For the first time in my life I was beginning to look critically at my role as an American in a foreign country and my responsibility for the actions of my country.

Despite what they were saying, I had no hatred for these men. I also, strangely, had no fear. Regardless of how the conversation sounded there was no personal animosity on either side. We were like two opposing sides of a sports team arguing over a recent match. These people were poor. They had nothing but their pride. They woke each day and felt the far-reaching hand of American foreign policy impacting their lives.

We continued to voice our disagreements in a mixture of English, Arabic, Italian, and French. Slowly, I was beginning to empathize with them. By the time the arguments on both sides had begun to repeat themselves, we were being treated to the sound of inking and stamping emanating from the office. That sound was music to the ear of any third-world traveler. If we had not made any progress toward world peace, at least progress was being made in our quest to get out of Djibouti.

This progress was short-lived. At the port gate an armed guard stopped the van and began an animated argument with our driver. It seemed that regardless of the validity of our papers our trip to the airport would be delayed until the appropriate departure tax had been paid. It was apparent that the tax in dispute was nothing more than baksheesh to line the pockets of the port police. Amy, having reached her personal limit for delay, walked to the front of the van.

"What does he want?" she asked the driver.

"He says he needs money before we can leave. I have told him to talk to my brother in the car behind us. He says he wants to talk to whoever is in charge."

"Tell him I'm in charge. All of our papers are in order and he will not get one dollar from me or anyone else in this van," Amy said before defiantly walking back to her seat.

As if not expecting to have to deal with an irate woman, the guard left the van to argue with the agent's assistant in the car behind us. After a few moments the guard stepped back in the van, took a seat in front, and with his automatic rifle lying across his lap, proceeded to direct the driver through the now-open gate.

"I don't know what was agreed to, but my grant is not paying a nickel for this," Amy said to me. Despite the delays we managed to arrive at the airport an hour prior to our flight's departure, still in the company of our armed escort.

Hordes of people blocked the single entrance to the terminal building as they awaited the arrival of Air France's only flight of the day. We gathered our duffels off the bus and pushed our way through the crowd. Amy struggled to keep one hand on my arm while managing to hold on to her bag. We made it to the open doorway and filed past two armed men in uniform. One of the men thrust out his hand and prevented Charlie from entering. After a brief negotiation that I couldn't hear over the din of voices, Charlie was allowed to join us.

The terminal was less congested than it had been outside, but the volume of noise and the temperature were considerably greater. The environment was made all the more uncomfortable by two barking dogs. It was not obvious whether the caged animals behind the ticket counter were arriving or departing, but if I could have spoken dog, I would have told them to be thankful they had no luggage to lose.

"I'm sorry, but we cannot put you on this flight," said the Air France agent. "All flights in and out of the United States are canceled indefinitely and it is airline policy not to board you if your connection is cancelled."

"Well, what are we supposed to do?" I asked.

"I would suggest staying here until things clear up in the States."

Amy and I gathered the rest of the group to see if everyone else was getting the same answer.

"The agent I talked to said we could get on the flight to Paris, but we'd be on our own. The airline will not take responsibility for taking care of us there," said Charlie.

"So here are our options: stay here and eat parasite-infested fish with warm goat's milk or get stuck in Paris for who knows how long, and survive on truffled foie gras and fine Bordeaux." I may have exaggerated the culinary conditions in Djibouti, but I made my point.

In a unanimous decision we bit the bullet and risked being stranded in Paris.

Airline policy of not allowing passengers to fly without a connecting flight remained a problem. The frantic agent at the counter was preoccupied with a host of other crises unrelated to us, and was not very interested in providing the details of a solution. He did, however, offer a promising option: we could sign a waiver of liability. The only catch was he didn't have any such document and he did not have the time to write one himself so it was up to us to write the terms of the waiver. He handed me a blank sheet of paper.

Using my best legalese, with a lot of "party of the first part" and so on, I constructed a document that any Harvard lawyer would have been proud of. We would not hold the airline accountable for our expenses while in France. And, because I didn't believe the agent would even read it, I signed the waiver with a flourish as "Daffy Q. Duck." As I suspected, with all the confusion surrounding the ticketing desk, the Djibouti agent did not even bother to look at my efforts. Instead, Amy and I were handed boarding passes and along with the rest of the group found our seats on the plane to Paris.

The flight to Paris was not without tension. A scheduled stop in Jeddah, Saudi Arabia, had us on edge. Most of the Djibouti passengers deplaned, but because we were continuing on to France, we were required to remain aboard. As on the flight to Djibouti, security personnel walked the isles searching for any sign of alcohol or contraband, but this time they were more diligent, as if they were looking for something more. Tensions between the United States and the Middle East had been growing for months and the events of the last few days made everyone uneasy. The guards inspecting the plane probably did not even know we were Americans, but as dozens of embarking passengers in white robes and red-checked head scarves took their seats, I felt as if I was wearing neon red, white, and blue. The flight itself proved uneventful, but we were not prepared for what we were to encounter once the plane landed.

At Charles de Gaulle airport there was chaos around the international ticket desks. A crowd 15 deep was vying for the attention of two agents. People were demanding to be placed on the first flight to the United States. They sounded like a pack of misogynistic assholes: "I must be in Houston this evening," a man screamed to a woman behind the counter who had probably heard every life-or-death reason a hundred times that day. There was a smattering of "do you know who I ams" and plenty of tearful pleas, but it was obvious to me that nobody was going anywhere in the near future.

After two hours I made it to what approximated the head of the line. In the most polite and self-deprecating voice I could muster I asked the tired woman on the other side of the desk what she thought was our best course of action. She looked at me for a minute and then at the rest of our group milling about beyond the crowd. "Wait one moment." She briefly stepped into the back room before returning. Quietly she said, "Go to this address. You'll need to take a taxi." She handed me a small yellow piece of paper with a handwritten address scribbled on it. I thanked her and discreetly pocketed the paper.

We left the angry crowd of fellow Americans behind, and squeezed our small group of tired, smelly, and shell-shocked bodies into three cabs. Charlie handed the driver the yellow note, and he nodded his understanding before getting out to confer with the other two taxis.

The drive took us through Paris, north, into suburbs and then, after almost an hour, the rural French countryside. None of us knew where we were going, or what to expect. At this point in our journey we were so exhausted, both physically and mentally, that we were entirely willing to leave ourselves at the mercy of others.

The cars entered a quaint village surrounded by fields of flowers and vegetables. Suddenly the driver turned sharply right, under a large stone arch. The taxi continued down a long red brick drive overhung with a canopy of branches from ancient elm trees.

"Shit," said Charlie, looking out of the window with an awed look on his face. "I think he's got the wrong address."

Through the windshield, I saw a circular drive that turned in front of an ivy-covered stone mansion. Flower boxes overflowing with brightly colored blossoms hung from three stories of diamond-paned, leaded-glass windows. As the car came to a stop a uniformed bellman approached.

"Are you sure this is the right place?" Amy asked the driver in French.

"Oui," he replied, pointing to the yellow paper.

At the check-in desk the attendant gave us each a key to our rooms. "I apologize, but your rooms are not quite ready. Feel free to wait in the lounge. Other than that, you are all set, Monsieur."

"But who arranged this, how much will it cost?" I asked.

"Don't worry, it has all been confirmed. Please let us know if there is anything at all we can do for you."

We looked blankly at one another. From the lounge came the familiar sound of American television, and as a group we were drawn to it. The CNN anchor was broadcasting into an empty room when we walked in. Images from New York taken two days before filled the screen. This was the first video we had seen of the attacks on the World Trade Center and the Pentagon. I watched transfixed. The images were more horrific than those I created in my mind. Amy sat close to the TV, trying to see the details of what had happened. Nobody spoke. It was as if we were reliving the moment all over again. We watched the news until it became sickeningly redundant and we realized that we hadn't eaten all morning.

"Where can we get something to eat?" I asked the bartender.

"The restaurant is not open yet, but I will ask," he answered in thickly accented English. He came back a moment later and gestured for us to go into an adjoining room.

The Maitre'd showed us to a large round table covered in a starched linen cloth. Crystal wine and water glasses complemented the formally set china and silverware. A waiter brought us each a small glass of something to drink. "Compliments of the hotel," he said. It was an incongruous scene; eight people in dirty tee shirts and shorts, who had not bathed in days, sitting in a Michelin-rated restaurant drinking champagne cocktails. As I looked down at my plate of duck a l'orange I gave thanks that it was not parasitic fish and sour goat's milk.

The glory of the first hot bath and wonderful food steadily gave way to the reality of being stuck far from home with no plan to get there. As each day passed and flight after flight was canceled, time in our little world seemed to stop. Even the good food and sights of France could not distract us from our situation.

To relieve the cabin fever beginning to fester within us, Amy and I rented a car to visit the Memorial de Caen, a museum dedicated to the history of World War II near Normandy. The approach to the entrance was a long series of concrete steps ending at a memorial to the nations that helped liberate France. A row of flagpoles lined one side, flying flags of the allied nations. Under the American flag sat a pile of flowers and cards. Many were printed in English: "It is our turn to help," "God bless America," and "we will never forget" were some that I read. The hospitality and comfort that the French people had shown to us since we'd arrived was not something I would forget either.

After almost a week, and hundreds of dollars worth of phone calls, we had secured seats on a flight to Boston. At the Paris airport we saw a man wearing an R.V. *Raven* tee shirt. It seemed so coincidental that I had to approach him. His name was Fernando and he was flying to the Seychelles to meet the ship for a four-month tour. Amy and I invited him to share a coffee with us. While the three of us waited for our respective flights we told Fernando of our experiences during the last cruise. He had heard about something happening, but didn't know any of the details. After hearing our story he had a lot of questions, but seemed unconcerned about finding his way to the ship. As things stood he did not know what to expect when he got there, just as we did not know what to expect when we got home. We only know that it would be better than any place we'd been so far.

The jumbo jet was at 35,000 feet, somewhere over the mid-Atlantic ridge. Amy turned to me and said, "So, do you think you want to go to sea again?"

I hadn't thought about it. Time and distance tend to erase the bad memories. "I have just completed the worst cruise in my short oceanographic career. Any others in my future can only be an improvement."

"What if we couldn't go together anymore?" Amy asked.

"Why would that be?"

"If we have children one of us would have to stay home."

"We never did have that discussion about kids." I smiled, thinking about when during that cruise would have been an appropriate time to talk about it.

"I think we should adopt," Amy said bluntly.

"I think we should too," I said, surprising myself at reaching the conclusion so quickly.

We had talked about adoption before, but like every other conversation about a family, it never became serious. Now things felt different. With all that was going on, children should have been the last thing on our minds, but for reasons I could not explain I wanted a family. Maybe it was a defiant act in the face of a society on the verge of collapse, or maybe it was the result of so many recent thoughts about my own mortality. Whatever the motivation, it seemed like right thing to do and the right time to do it.

"It will be a big change," Amy said.

"I'm ready for a change."

Long ago, when the subject of having biological children had come up, we were both drawn to the concept of adoption. It wasn't for the reasons people would have guessed: passing on some blindness genetic trait, or our advancing age, or even that fertility was a problem. The decision to use adoption was as simple as "it felt right." The *how* we would create a family was easy. What had proved difficult was the *when.*

Like the fertilization of egg with sperm, in that plane over the Atlantic, the how and the when came together.

The Author

David Fisichella manages shipboard scientific services at the Woods Hole Oceanographic Institution. Formerly an engineer with an aerospace company, he turned to providing adaptive technology to the blind. When his first marriage collapsed, he went to sea on a whim with a scientist suffering from macular degeneration. David teaches sailing and skiing to the blind, and sails around New England with his wife and 8-year-old daughter.

About the Type

This book was set in ITC New Baskerville, a typeface based on the types of John Baskerville (1706-1775), an accomplished printer from Birmingham, England. The excellent quality of his printing influenced such famous printers as Didot in France and Bodoni in Italy. Baskerville produced a masterpiece folio Bible for Cambridge University, and today, his types are considered to be fine representations of eighteenth century rationalism and neoclassicism. This ITC New Baskerville was designed by Matthew Carter and John Quaranda in 1978.

Designed by John Taylor-Convery
Composed at JTC Imagineering, Santa Maria,CA